M000096987

rder pour
kersfield Memorial Stadium
ntractor: Tumblin Company

Edward J. Monahan
CONSTRUCTION OF AND ON

Ben C. Gerwick, Jr.
CONSTRUCTION OF OFFSHORE STRUCTURES

David M. Greer and William S. Gardner
CONSTRUCTION OF DRILLED PIER FOUNDATIONS

James E. Clyde
CONSTRUCTION FOREMAN'S JOB GUIDE

Leo Diamant
CONSTRUCTION ESTIMATING FOR GENERAL CONTRACTORS

Richard G. Ahlvin and Vernon Allen Smoots
CONSTRUCTION GUIDE FOR SOILS AND FOUNDATIONS,
Second Edition

B. Austin Barry
CONSTRUCTION MEASUREMENTS, Second Edition

John E. Traister and Paul Rosenberg
CONSTRUCTION ELECTRICAL CONTRACTING, Second Edition

Harold J. Rosen and Tom Heineman
CONSTRUCTION SPECIFICATIONS WRITING: PRINCIPLES AND
PROCEDURES, Third Edition

Leo Diamant and C. R. Tumblin
CONSTRUCTION COST ESTIMATES, Second Edition

Thomas C. Schleifer
CONSTRUCTION CONTRACTORS' SURVIVAL GUIDE

A. C. Houlsby
CONSTRUCTION AND DESIGN OF CEMENT GROUTING: A
GUIDE TO GROUTING IN ROCK FOUNDATIONS

CONSTRUCTION
COST ESTIMATES

CONSTRUCTION COST ESTIMATES

SECOND EDITION

LEO DIAMANT, B.C.E.
C. R. TUMBLIN, P. E.

WILEY

A WILEY-INTERSCIENCE PUBLICATION

JOHN WILEY & SONS

New York · Chichester · Brisbane · Toronto · Singapore

Copyright © 1990 by John Wiley & Sons, Inc.

All rights reserved. Published simultaneously in Canada.

Reproduction or translation of any part of this work
beyond that permitted by Section 107 or 108 of the
1976 United States Copyright Act without the permission
of the copyright owner is unlawful. Requests for
permission or further information should be addressed to
the Permissions Department, John Wiley & Sons, Inc.

Library of Congress Cataloging in Publication Data:

Diamant, Leo, 1920–
    Construction cost estimates / Leo Diamant, C.R. Tumblin. — 2nd
ed.
      p.   cm. — (Wiley series of practical construction guides)
    Rev. ed. of: Construction cost estimates / C.R. Tumblin.
    "A Wiley-Interscience publication."
    Bibliography: p.
    Includes index.
    ISBN 0-471-60020-2
    1. Building—Estimates. I. Tumblin, C. R., 1917–
II. Tumblin, C. R., 1917–   Construction cost estimates.
III. Title. IV. Series.
TH435.D433   1990
692'.5—dc20   89-16483
                                        CIP

Printed in the United States of America

10 9 8 7 6 5 4 3 2 1

Congratulations! You've just bought a profit-making tool that is inexpensive and requires no maintenance, no overhead, and no amortization. Actually, it will increase in value for you each time you use this volume in the Wiley Series of Practical Construction Guides. This book should contribute toward getting your project done under budget, ahead of schedule, and out of court.

For nearly a quarter of a century, over 50 books on various aspects of construction and contracting have appeared in this series. If one is still valid, it is "updated" to stay on the cutting edge. If it ceases to serve, it goes out of print. Thus you get the most advanced construction practice and technology information available from experts who use it on the job.

The Associated General Contractors of America (AGC) statistician advises that the construction industry now represents close to 10% of the gross national product (GNP), some 410 billion dollars worth per year. Therefore, simple, off-the-shelf books won't work. The construction industry is unique in that it is the only one where the factory goes out to the buyer at the point of sale. The constructor takes more than the normal risk in operating a needed service business.

Until the advent of the series, various single books (many by professors), magazine articles, and vendors' literature constituted the total source of information for builders. To fill this need, this series has provided solid usable information and data for and by working constructors. This has increased the contractors' earning capacity while giving the owner a better product. Profit is not a dirty word. The Wiley Series of Practical Construction Guides is dedicated to that cause.

M. D. MORRIS, P.E.

Ithaca, New York
November 1989

v

PREFACE

In the second edition, we have enlarged and broadened the scope of trades estimated.

The first edition included numerous examples of concrete structures and trades primarily involved in engineering works (heavy construction). We have expanded the book to include examples of estimating the many trades in building construction (light construction).

The revised edition has been enhanced better to serve the field construction workers and job superintendents whose ambition is to start their own general contracting business or move into the general contractor's home office as an estimator or project manager.

Although this book stands on its own, in conjunction with *Construction Estimating for the General Contractor* by Leo Diamant (Wiley Series of Practical Construction Guides), all that is required for estimating and bidding construction projects is thoroughly covered.

LEO DIAMANT, B.C.E.

Valley Stream, New York
January, 1989

PREFACE TO FIRST EDITION

This book presents a practical method for preparing consistent and detailed cost estimates for construction projects. The writer is a general contractor with more than 35 years of successful business experience bidding and constructing engineering and building projects. The procedures illustrated are the ones used, with minor variations, throughout the industry.

Although the material is written primarily for upper division students in engineering or architecture who plan to enter the construction industry, it is also intended as a reference book for general contractors and for those professionals who are involved with the preparation of job plans and specifications and with the awarding of construction contracts.

Every construction project consists of a number of separate categories of work called job items. The estimating process requires each job item to be analyzed, divided, and subdivided, as necessary, to arrive at those primary activities that must be performed to complete the various operations involved.

The quantities of the primary activities and materials required are collected and summarized on the takeoff for the item. The estimated cost for the item is then determined by applying a system of unit costs to these quantities. The unit costs for labor and equipment are selected from an array of unit costs or production rates achieved on past jobs involving operations similar to the ones being considered.

Each contractor's unit costs for labor and equipment are unique and reflect his efficiency and experience, the job conditions relating to his cost data, the amount of job supervision used, and the quality of his cost-keeping system.

The recognition and the quantification of the required primary activities is of the utmost importance since they must be priced to estimate job costs, timed to estimate the time required for completion, and monitored to control the cost and the progress of work under construction.

In the following chapters, typical job items are analyzed and the primary activities relating to the various types of work are identified. Takeoffs are made illustrating the methods used to collect and summarize the quantities involved. Unit costs are presented, and these unit costs are indexed for conversion to those wage and rental rates applicable to the job being bid. Cost estimates as well as estimates of the time required for completion are illustrated for various job items and for assumed projects consisting of various combinations of job items. The estimates are identical to those made in practice; the job cost estimates include the costs of direct job overhead, general overhead, payroll expenses, and profit.

The unit costs presented herein cover those job items normally performed by general contractors and are quite inclusive. The role of the subcontractor is discussed in detail, but unit costs for subcontract work are not covered. In the job cost estimates illustrated, assumed quotations from subcontractors are used in conjunction with the cost estimates made for those items of work normally performed by the general contractor.

The material is presented so that by the time you have progressed through part of the material covered in Chapter 8, you will be able to prepare "dry-run" bids for small construction projects, and these bids can be compared with the actual bids submitted by the competing contractors. Such an exercise will demonstrate your proficiency in applying the methods illustrated, but your estimate cannot be considered reliable or sound since field and office experience are required to properly modify average unit costs to suit particular job conditions. In addition, such exercises will not furnish any indication of your ability to actually do the work at the estimated cost, and you will not experience the tension and excitement that prevail on bid day, when the bid is backed with money and submitted against an absolute time deadline.

The ability to analyze the work going on about them will make the early years of on-the-job apprenticeship much more productive for those entering the construction industry, both in terms of their own professional development and in terms of their value to and rate of advancement by their employers. It will also enable them to start a practical file of job unit costs based on their own job experiences. A file of such unit costs is a basic necessity for all construction cost estimators.

Those who enter the construction industry will discover a world of unlimited opportunity. With each job well done, they will experience that quiet pride and inner satisfaction known only to those who build enduring things; and when their days in construction are over, their works will stand in silent tribute to their early struggles, to their efforts, to their abilities, and to their creativity.

C. R. TUMBLIN

Bakersfield, California
July 1979

CONTENTS

TABLES

ABBREVIATIONS

A	Area
AB	Aggregate base
AS	Aggregate subbase
BCY	Bank cubic yards
BF	Base frame or base frames; backfill
BF & E	Base frame or frames and extension frame. Thus 2 BF & E refers to a tower 2 base frames plus an extension frame in height.
C	Carpenter
CF	Cubic foot
CIDH	Cast-in-drilled-hole
CJ	Construction joint
CTB	Cement-treated base
CWT	Hundredweight
CY	Cubic yard
DE	Deck edge (square feet)
DF	Douglas fir
DO	Direct overhead
F	Cement finisher or finish
FG	Fine grade
FICA	Federal Insurance Contributions Act (social security)
FIN FLR	Finish floor
FOB	Freight on board
FUI	Federal unemployment insurance
$H = 5$	Height of form panels—in this case, 5 feet
$H_c = 18$	Height to concrete soffit, measured from ground line—in this case, 18 feet

HEx	Hand excavation
HP	Hinge point, horse power
K	Cost index (see Chapter 2) such as
K_F	Hourly wage rate for one cement finisher
K_L	Hourly wage rate for one laborer
K_{L+F}	Hourly wage rate for one laborer plus one cement finisher
K_{6+2}	Total wages for six carpenters plus two laborers working for 8 hours
K_{Cr}	Hourly rental rate for a 35-ton truck crane
K_{BH}	Hourly rental rate for a small rubber-tired loader-backhoe
L	Length, or laborer, or labor
LS	Lump sum
LCY	Loose cubic yards
LF	Linear feet
M	Materials, or thousand
MB	Machine bolt
MC	Medium curing
MCY	Thousand cubic yards
MFBM	Thousand feet, board measure
M, R	Material or rental costs
MSF	Thousand square feet
MSY	Thousand square yards
NOC	Not otherwise classified
NTS	Not to scale
OD	Outside diameter
OG	Original ground surface
OH	Deck overhang, or overhead
P	Pour or place concrete
PSF	Pounds per square foot
psi	Pounds per square inch
R	Rentals or rough
RC	Rapid curing
SC	Slow curing
SF	Square feet
SS	Set and strip
SS, P & F	Set, strip, pour and finish
STA	Stations
SUI	State unemployment insurance
SY	Station yards or square yards
S1E	Surfaced one edge
S1S1E	Surfaced one side, one edge
T	Tons (2000 pounds)
t	Thickness
T & G	Tongue and groove
W	Width

WF	Wall form
WW	Wingwalls
Y	Yards
2×, or 4×	Read 2 by, or 4 by. In carpentry, this refers to material which is 2 or 4 inches in thickness

CONSTRUCTION COST ESTIMATES

Basic Construction Terminology and Procedures

Field experience on construction projects is absolutely essential for those students planning to become general contractors or construction cost estimators. Ideally, this experience should be gained while working in a supervisory capacity for a general contractor. Every construction cost estimator must have the background and ability to function comfortably as superintendent on any job he bids.

This book presents a practical method for preparing detailed cost estimates for construction projects. Since it is written for students who normally will not have had any construction experience and those with little knowledge of the workings of the general contractor's home office, this chapter is devoted to the general aspects of the contract procedure and to the elements involved in the bidding and the awarding of construction projects. An understanding of these processes is a necessary prerequisite to the preparation of construction cost estimates and will also benefit those young engineers and architects engaged in pure design work who will frequently interface with general contractors during the course of their professional careers.

1.1 DEFINITIONS AND TERMS

An understanding of the following terms is requisite to the discussion of construction cost estimates.

Bare Cost. The bare cost of a job includes all costs except the contractor's allowance for general overhead and profit. The bare cost of a job item includes the costs for the labor, materials, and equipment required to complete the item, and the labor cost used will include the contractor's allowance for payroll expenses.

Bid Price. The bid price for a construction project is a firm price for completing the work in accordance with the contract documents. It is submitted to the owner, in writing, by the general contractor. The bid price is determined on the estimate summary sheet and is based on detailed cost estimates. All risks relating to unforeseen job conditions or resulting from estimating errors or omissions are borne by the contractor.

Estimate Sheet. The sheet on which the estimate for the bare cost of a particular job item is made. Typically, sheets from standard $8\frac{1}{2} \times 14$ columnar pads are used.

Estimate Summary Sheet. The sheet on which the bare costs for all of the job items are tabulated together with the costs for direct job overhead, general overhead, and profit. The total of these costs is the bid price for the job. Typically, sheets from standard $8\frac{1}{2} \times 14$ columnar pads having line numbers at the left margin are used.

Item. The meaning of this term depends on the context in which it is used. A job item is a separate category of work on a particular project. A major item is any category of work which constitutes a large portion of the overall job cost and may be a job item or a portion of a job item.

Primary Activity. As used herein, a primary activity is a fundamental operation necessary to complete a portion of a job item. The unit cost for performing a primary activity includes the costs of those secondary activities which must normally be performed to complete the operation. The primary activities are so selected that the relative amounts of these secondary activities will be reasonably constant from one job to another.

Production Rate. Production rates are expressed in units of work accomplished per hour for a given crew or piece of equipment. Unit costs may be obtained by dividing the hourly cost of the crew or equipment by the production rate. Production rates determined from past job cost records are said to be "effective rates" since they reflect the inevitable minor delays and interruptions that occur on all jobs.

Rental Rates. The rental rates for motorized equipment include the costs of operation and maintenance and the costs of subsistence payments to the operators when required.

Unit Costs. Unit costs are expressed in dollars per unit of work performed and are used extensively in the preparation of construction cost estimates. All unit costs must be indexed to the wage or rental rates on which they are based. A production rate may be obtained by dividing the hourly cost of the proposed crew or equipment by the unit cost.

Unit Price.   The unit price is the price bid for completing an item on a unit price contract. The total bid price for a unit price contract is the sum of the products of all the quantities for the various job items and their respective unit prices.

Wage Rates.   The wage rates, to which unit costs for labor are indexed, include the base rate of pay plus all fringe benefits.

1.2   THE ESTIMATING PROCESS

The estimating process construction cost estimates for the general contractor consist of the total cost of all the bids of the subcontractor and the total cost of the work constructed by the general contractor's own forces. The general contractor engaged in building construction (light construction) will do most of the work with subcontractors. The general contractor's estimating staff will take off, in addition to the work performed by his own force, a number of the key trades performed by his subcontractor. Takeoff of the subcontractor's work is desirable (1) for checking the accuracy of the subcontractor's takeoff and (2) as a tool in negotiating for the lowest bid.

General contractors in engineering work (heavy construction) will do most of their work with their own forces.

The soundness of the completed estimate depends on the following three factors:

1. The accuracy of the takeoff.
2. The accuracy of the extensions, additions, and transcriptions.
3. The judicious selection of the unit costs and production rates to be used.

1.2.1   Elimination of Errors

Each takeoff is made to the required degree of precision, and all work is checked. As illustrated in Chapter 8, the primary activities required to complete each item are tabulated and collected in an orderly and systematic manner.

All errors in arithmetic and errors of transcription must be eliminated. All extensions are checked and, where the quantities are large, double checked. All totals are double checked and care is taken to ensure that all figures are clear and legible.

As each step is completed, the work is given a final visual check to eliminate possible errors in magnitude or transcription. The estimate sheets are checked against the takeoffs to ensure that no quantities have been overlooked. The estimate summary sheet is checked to ensure that all trades or job items have been listed; and on unit price contracts the quantities listed on the summary sheet for each job item are checked against those listed on the bid proposal.

As the estimate for each job item is completed, the overall unit cost for the item, as estimated, if possible is compared with the overall unit costs realized on past jobs for similar types of work. When significant variation is found, its causes are investigated; this process is simplified when cost records are available to show the overall unit costs for the typical components of similar items of work.

1.2.2 Unfamiliar Minor Items

Estimators are frequently confronted with unfamiliar minor items for which no cost data are available. In such cases, unit costs are reviewed for familiar operations which may or may not be similar to the one in question. A unit cost is selected for a familiar operation that, in the estimator's judgment, will cost more than the one in question, and a second unit cost is selected for an operation that should cost less.

With the probable unit cost for the item in question bracketed, a trial unit cost is selected and converted to a production rate. This rate is considered in view of the anticipated job conditions and modified if necessary, and the resulting unit cost is entered on the estimate sheet.

On occasion, all estimators face an approaching bid deadline but lack a quoted price for some minor piece of equipment. At any given time, however, the cost per pound is relatively constant for similar types of equipment. A rough estimate can be made in such cases on a weight basis if the current cost per pound of another piece of somewhat similar equipment is known.

Relative weights are also useful when last-minute quotations for equipment "in lieu of that specified" must be considered. It is always doubtful that a substitution would be permitted where the substitute equipment is appreciably lighter than the equipment called for in the specifications.

1.2.3 The Estimator and the Job Superintendent

Some job superintendents and some crew foremen will do work at lower unit costs than others. The productivity of the proposed job personnel is always considered by the estimator during the unit cost selection process.

Prior to starting a new job, the estimator confers with the superintendent regarding the job procedures and schedules on which the estimate was based. The job superintendent is responsible for completing the work at or below its estimated cost, and in some cases procedural changes may be made. During the course of the work, the estimator monitors the developing unit costs and confers with the superintendent on a regular basis.

1.2.4 Selection of Equipment

All construction cost estimates are based on using particular types of equipment. Normally, equipment selection presents no problems. The contractor will be bidding familiar work in a familiar area, and the choice of equipment will be obvious.

On large or unusual jobs, highly specialized and, in some cases, specially built equipment may be required. In such cases the final selection of equipment is made by management, and the production rates for the equipment may be based on the specifications for the equipment together with other information supplied by the manufacturer.

1.3 THE CONTRACT PROCEDURE

On every construction project there are but two primary parties or principals involved: the owner, who wants the job done, and the general contractor, who agrees to build it for an agreed price in a specified period of time.

1.3.1 The Job Architect or Engineer

Upon deciding to have a facility built, the owner engages the services of an engineer or architect who prepares the contract documents. Engineers are normally engaged by the owner when the proposed project consists primarily of engineering work such as roads, canals, dams, bridges, and certain industrial facilities. Architects are usually engaged when the proposed project consists primarily of buildings. On most large projects the architect will, in turn, engage the services of a structural engineer, a mechanical engineer, and an electrical engineer, who report to and are paid by the architect.

In all cases the contract documents will consist of a set of plans, the job specifications, the bid proposal, and the contract agreement. In addition, a notice to contractors, contract bonds, and a bidder's bond may be included. While the plans are being prepared, the architect or engineer maintains close contact with the owner and furnishes him with preliminary cost estimates for the developing project. These estimates enable the owner to confirm the economic feasibility of the project, and the plans may be modified to fit the project to the funds available.

Upon his final approval of the plans and specifications, the owner selects a contractor to build the facility. This may be done through negotiations with one or more contractors, or a number of responsible contractors may be invited to bid the project, which will subsequently be awarded to the lowest bidder.

Skepticism is sometimes expressed regarding the concept of dealing with the lowest bidder. Such skepticism may be warranted in those cases where the prices of products or services are compared without regard to the specifications to which the products have been manufactured or to the experience and integrity of the suppliers of the services.

In the construction industry it is the responsibility of the architect or engineer to furnish a complete and detailed set of plans and specifications and a qualified job inspector or resident engineer and to limit the bidding to competent and responsible contractors. Under such conditions, the awarding of the job to the lowest bidder will enable the owner to obtain the exact job desired at the lowest possible price. From the owner's point of view, his selection of the architect or engineer is of the utmost importance, and he should exercise great care and his best judgment in making this choice.

1.3.2 Notice to Contractors

When the job is to be awarded on the basis of competitive bids, the owner sets a bid date together with a place and time for the bid opening, and the architect or engineer notifies the contractors that the job is out for bids. Public agencies have strict, legislated procedures for notifying contractors, and these procedures must be followed in detail on such work. In practice, all contractors subscribe to daily construction services which publish the essential information concerning jobs out for bid as well as the prices subsequently bid on all projects in any given geographical area.

1.3.3 Prequalification of Bidders

To ensure that all contractors bidding their work are competent and experienced, many public agencies require all prospective bidders to be prequalified. In such cases the contractors files a financial statement and an experience record with the agency on a yearly basis and is given a rating expressed in terms of the maximum total dollar volume of work he will be allowed to have under construction with the agency during the course of the year. The required bid proposal form will not be issued to any contractor if the estimated cost of the proposed project plus the value of other work he may still have to complete for the same agency exceeds his prequalification limit.

Prequalification of contractors is an excellent requirement from all points of view. It enables the owner to accept the lowest bid with confidence, and it provides an impartial and fair method of screening out prospective bidders who are too small or too inexperienced to bid on a particular project. This screening process thus provides definite benefits to the owner, his engineer or architect, and those contractors who are screened out.

Prequalification should be required more often by architects and engineers in connection with larger private work. The prequalification limit in such cases could be based on a copy of the contractor's current letter of prequalification from the appropriate state agency, and no further analysis by the architect or engineer would be required.

1.3.4 The Cost Estimate

Upon receiving the plans, specifications, and bid proposal, each general contractor prepares a detailed cost estimate for the job. Time is often critical in the preparation of such estimates. The contractor may have only 5 or 6 weeks to bid a five-story concrete building occupying a city block, for which the concrete takeoff alone may require as much as 160 man hours of work.

During the preparation of the cost estimate, the estimator must not spend time analyzing unnecessary details. Each job item must be analyzed only to the degree necessary to determine the lowest probable cost for the job as a whole. Extreme care must be used in estimating the cost of major job items, but time must not be wasted in efforts to be too precise in determining the costs of insignificant items.

1.3.5 Subcontract Costs

In addition to the cost of the general contractor's own work, estimate will include the costs of a number of subcontractors. During the course of construction, the subcontractors report to, are under the general direction of, and are paid by the general contractor. Subcontractors are often referred to by the trade or specialty they perform. Thus the painting subcontractor will frequently be called the painting contractor. The term ''contractor'' by itself normally refers to the general contractor. On building projects, about 75 percent of the work is done by subcontractors, while on engineering work the general contractor will do about 65 percent of the work himself.

In all cases the general contractor has the sole responsibility for the successful completion of the project. He furnishes the overall job supervision, prepares all time schedules, and coordinates the work of all of the subcontractors with his own work to ensure that the project is completed in strict accordance with the plans and specifications, in the least amount of time, and at the lowest possible cost to the owner.

1.3.6 The Bid Opening

On the bid day all bids are submitted in sealed envelopes at the time and place designated, and no bids are accepted after the specified time for the bid opening. The bids are normally opened in public at the time specified and are taken ''under advisement'' for a few days to enable the owner and his architect or engineer to carefully check and consider each proposal. Most specifications allow the owner to reject any or all bids, and all bids are rejected on rare occasions. In the typical case the job is awarded to the lowest bidder.

1.3.7 The Contract Award

Upon being awarded the job, the contractor and his bonding company sign the contract bonds and deliver them to the owner, and the owner and the contractor sign the construction contract. The contractor must then commence work within a specified period of time, which frequently will be 5 to 10 working days.

After the contract is signed, the owner is represented by the architect or engineer in all further dealings with the contractor. The contractor does the work in accordance with the plans and specifications and to the satisfaction of the architect or engineer. At the job level, the contractor's superintendent deals on a day-to-day basis with the architect's representative or job inspector or with the resident engineer in the case of engineering work.

1.3.8 Pay Estimates

During the course of the work the contractor is paid monthly by the owner on the basis of monthly pay estimates which are approved by the architect or engineer.

On most jobs the owner retains a specified amount, usually 10 percent of the total amount earned, during the course of the job and for a specified period of time after completion. The retention period after completion is specified to allow mechanics' liens and claims from material or equipment suppliers to surface, if any are forthcoming. In practice, the contractor will receive the final payment about 6 weeks after the owner has accepted the work.

In most cases the contractor receives the monthly payment from the owner 2 to 3 weeks after the monthly pay estimate has been submitted for approval. Because of this time lag and the retention, it is important for the contractor to exercise sound judgment in determining how he will distribute his allowances for direct job overhead, general overhead, and profit among the job items on engineering work or among the sections of the cost breakdown on building work.

1.3.9 Contract Change Orders

Unforeseen changes in the proposed work, which become necessary or desirable during the course of the job, are specified and paid for on contract change orders. Change orders are negotiated at the job level, approved by the architect or engineer, and paid for at agreed prices when possible. In cases where it is difficult to arrive at an agreed price, the change order work may be done on a time and material or force account basis. In such cases, the contractor does the required work and is paid for his actual costs plus a percentage for overhead and profit, as set forth in the specifications. If the change order work falls on the critical path for the project, a time extension is negotiated and becomes a part of the contract change order.

1.3.10 Liquidated Damages

Specifications may require the contractor to pay liquidated damages to the owner for each calendar day that the actual time required for completion exceeds the time specified for completion. Such damages may vary from $100 a day for a simple box culvert job to several thousand dollars a day on large contracts.

Liquidated damages represent an attempt to arrive at a predetermined and agreed price for the actual damages that will be suffered by the owner in the event of delays in completing the work. From the owner's point of view, it is legally important that the specifications do not refer to liquidated damages as a penalty.

When liquidated damages are specified, it is essential that the specified time for completion be reasonable. When, due to unusual circumstances, the time specified for completion is based on using double shifts, this should be set forth in the specifications.

Liquidated damages will not ensure that an impossible time schedule will be met. In such cases the estimated cost of the liquidated damages, based on a realistic time schedule, is added to the estimated bare cost of the job as an added item of expense.

1.4 THE CONTRACT DOCUMENTS

The contract documents must include the plans, specifications, bid proposal, and contract agreement. A notice to contractors, contract bonds, and a bidders bond may be included. All other contract documents are included in the contract agreement by reference and are an integral part of the construction contract.

1.4.1 Plans and Specifications

On fixed price contracts, the owner must furnish a complete and detailed set of plans and specifications to the contractor. The contractors must be able to determine exactly what is required to be done and to be furnished. Ambiguous or incomplete plans and specifications require the contractors to make assumptions as to what may be required and often result in higher bid prices from the experienced contractors.

A standard set of specifications is frequently furnished together with a set of special provisions which relate the general requirements of the "standards" to the particular job by reference. On building work, the specials should include a separate section for each trade involved, and on engineering work the specials should include a separate section for each item of work.

The writer of the special provisions must be thoroughly familiar with the contents and requirements of the standards. Thus it is not sufficient for the specials to specify that the poles for the electroliers be in accordance with the standard specifications when the standards cover the requirements for painted as well as galvanized poles.

When it is desirable to change the requirements of the standards in connection with a given item of work, this is done in the specials by a specific reference. Thus, the specials might state that "in lieu of the method of payment set forth in section 63.07 of the standard specifications, the excavation will be paid for on a lump sum basis." In this case the specials would then say exactly what items were to be included in the lump sum paid.

Care must be exercised not to duplicate any of the requirements of the standards in the special provisions. Such duplication will inevitably result in conflicting requirements which will necessitate the issuance of addenda to the specifications during the time that the job is being bid. Discrepancies between the standards and the specials reflect careless specification writing and should not occur in specifications written by competent engineers or architects.

1.4.2 The Bid Proposal

The bid proposal is addressed to the owner and is signed by the contractor. The contractor inserts his bid prices into blank spaces provided, and no alterations or additions are allowed. A good bid proposal form should not require any price to be entered in more than one place, nor any price to be written in words.

Following the bid opening, all the extensions and additions on unit price contracts are checked by the owner's engineer. Where an extension is incorrect, the unit price is considered to be correct, and the extension, together with the total price bid, is corrected by the engineer to determine the true bid price.

On some building jobs, a list of proposed subcontractors must be submitted with the bid proposal, and the successful bidder can only use the listed subcontractors on the proposed project. During the time that quotations from subcontractors are being received, the list of subcontractors is kept current and in phase with the low quotations being used in the estimate. The requirement for the list of subcontractors results from the efforts of subcontractor associations to combat ''bid peddling'' by unscrupulous general contractors together with unscrupulous subcontractors. Top quality general contractors maintain strict ethics in this regard and will never reveal any subcontractor's quoted price.

1.4.2.1 ALTERNATES. Alternate bids are frequently required on building work, and minor alternates, involving only a few subcontractors or the general contractor's own work, can be processed accurately by the contractor on bid day. Such an alternate might be one requiring the additional cost for using a different type of lighting fixture in certain rooms in lieu of the ones specified in the base bid or the reduction in cost if certain planters specified in the base bid are deleted from the contract.

There is insufficient time available on bid day to accurately process a major alternate involving many or all of the subcontractors. The reasons for this shortage of time are inherent in the bidding process and are completely beyond the control of the general contractor. The problems involved are discussed in Section 1.9, which covers the procedure on bid day.

Bid proposals requiring major alternates will be submitted by the contractors since their livelihoods depend on their obtaining work, but as a rule the cost to the owner for such work will be greater than if the alternate were included as a part of the base bid. On building work, the spread between the base bids submitted by the first and second bidders may be 1 percent or less, while the spread between their major alternates frequently amounts to 20 percent or more.

Major alternates often cause the listed subcontractors to change depending on which combination of base bid plus alternates is selected by the owner. For the same reason, even the low general contractor may be indeterminate, and the owner will, in effect, have two low bidders. In such cases, the owner and the architect will have at least one unhappy contractor regardless of what decision they make.

The failure of such a proposal to present the owner with a single undisputed low bidder may result in considerable embarrassment and disadvantage to the owner in cases where other business or political considerations are involved. A competent architect performs a very real service for the owner when the need for major alternates is eliminated through proper use of preliminary cost estimates and discussions during the planning stage of the project.

1.4.3 The Bidder's Bond

The bidder's bond is a guarantee that the contractor will enter into the contract with the owner at his bid price if he is awarded the job. Such bonds are addressed to the owner, signed by the contractor and by his bonding company, and are normally in the amount of 10 percent of the amount bid. Should the contractor refuse to enter into the contract, he could be required to forfeit the amount specified on the bond to the owner.

The cost of the bidder's bond is nominal and is a general overhead item. Many bonding companies make no charge for supplying bidder's bonds to those contractors for whom they regularly furnish contract bonds.

1.4.4 The Contract Bonds

Two contract bonds are normally required. The labor and materials bond is a guarantee that the contractor will pay his workmen and his suppliers. The performance bond is a guarantee that the contractor will complete the work in accordance with the requirements of the plans, specifications, and contract. These bonds are addressed to the owner and are signed by the general contractor and by his bonding company.

In the event of default by the contractor, the bonding company will have the work completed, usually by another contractor. The bonding company will pay all outstanding bills, which are subject to lien laws, with the exception of insurance premiums and fees for accountants and attorneys. The bonding company will recover its costs by liquidating the contractor's business and personal assets, if necessary.

From the owner's point of view, it is essential that the specifications clearly define what conditions will constitute default by the contractor and what will be grounds for terminating the contract. In addition, the procedures and time periods to be followed and allowed in the event of default should be set forth clearly.

Some architects and engineers have experienced difficulty in getting a bonding company to take over a job on which the contractor was not performing. In most cases, such difficulties result from specifications which fail to precisely define the grounds and the procedures for terminating the contract.

The cost for the two contract bonds is obtained from the bonding company, and the cost will vary with the size and classification of the job and the performance record of the general contractor.

1.5 TYPES OF ESTIMATES

The term ''estimate'' is loosely used in the construction industry, and its precise meaning must often be determined from the context in which it is used. It may refer to detailed cost estimates, preliminary estimates, estimates of the quantities involved in the job items, or estimates of moneys due to the contractor in payment for work done.

1.5.1 Detailed Cost Estimates

These are the estimates prepared by contractors in bidding jobs, and such estimates are covered in depth in the following chapters.

All detailed cost estimates are based on a takeoff of the materials and of the primary activities required for each job item, and the quality of the completed estimate depends on the accuracy of these takeoffs.

The total cost of any project consists of four categories of expenses. These are the costs of those materials to be furnished by the general contractor, the costs of subcontractors, the general job costs, and the costs of the labor and equipment required to do the general contractor's own work. The costs of materials and the subcontract costs are based on fixed quoted prices and can be determined accurately. Although sound judgment is required, the general job costs can be estimated with a reasonable degree of accuracy. The area of greatest uncertainty in the preparation of detailed cost estimates lies in the selection of the proper unit costs or production rates for labor and for equipment in connection with the general contractor's own work.

The total labor and equipment costs for the general contractor's own job items consist of the sum of the products of many quantities and their respective unit costs. Although the majority of the unit costs will be about right, some will be too high and others will be too low. In practice the sum of these products, however, will be very close to the costs actually realized in those cases where experienced estimators use sound judgment in selecting unit costs based on past cost records. Extreme care must always be used in selecting unit costs where large quantities are involved. The two primary pitfalls to be avoided are the failure of an inexperienced estimator to recognize some essential activity and to overlook it entirely and where, in the case of unit costs based directly on production rates, some circumstance on the particular job that will greatly affect the rate of production goes unnoticed.

1.5.2 Preliminary Cost Estimates

Preliminary estimates are frequently referred to as architect's estimates or as engineer's estimates. Architect's estimates are often based on the total square feet of floor area or on the total cubic feet of building volume for a given type of building and a given class of building use. There are publications available that present such data together with the realized unit costs for buildings recently bid in various sections of the country. Such data are often used to supplement the architect's own cost records based on his own work in his own area.

Engineering projects are normally bid on a unit price basis. Daily construction services publish the results of all public bid openings in their areas, and a file of average unit prices for the various items of work can readily be built up. Contractors often make estimates based on such average unit prices and a published engineer's estimate of quantities, prior to having any plans or specifications, as aids in determining whether or not to bid a particular project.

Preliminary estimates can be made quickly but are not sufficiently accurate to use as the basis for a bid. The architect or engineer may, to some degree, risk his reputation with a preliminary estimate, but the contractor is risking money.

1.5.3   Quantities Estimates

The list of job items together with their estimated quantities is called the ''engineer's estimate'' and constitutes the major portion of the bid proposal form on engineering projects. In addition, the engineer's estimate should be the first page of the special provisions for engineering work.

The quantities shown on the engineer's estimate are for estimating purposes only, and the final pay quantities are based on calculations using plan dimensions supplemented by field measurements when necessary. The pay limits used in preparing the engineer's estimate are set forth in the specifications, and these are the limits used in determining pay quantities. The specified pay limits may be different from those required to do the work. The horizontal pay limits for structure excavation and structure backfill are frequently specified to be vertical planes, and the contractor may actually move and replace more than twice the number of cubic yards shown on the engineer's estimate.

The number of items on the engineer's estimate may vary from 4 or 5 in the case of a simple box culvert job to as many as 150 or 200 on a 15-mile stretch of freeway.

1.5.4   Pay Estimates

Pay estimates are estimates of the money due to and earned by the contractor. Monthly pay estimates are normally made during the course of the work and upon completion of the job. After completion and upon the expiration of the specified waiting period, the final payment is made based on the final pay estimate. The final payment consists primarily of the money retained during the course of the work and in the case of unit price bids is based on a final and precise determination of the total quantities of all of the job items.

On building work, which is bid on a lump sum basis, the contractor furnishes the architect a cost breakdown by job sections at the start of the work. Subsequent pay estimates are based on an estimated percentage of completion for the items on the cost breakdown.

1.6   TYPES OF CONSTRUCTION CONTRACTS

Negotiated contracts are frequently awarded on a cost-plus basis, while competitively bid contracts are always awarded on the basis of a fixed price for doing a specified job.

1.6.1 Cost-Plus Contracts

Cost-plus contracts are typically used in those cases where complete plans and specifications are not available at the time the owner must start the project. Under the cost-plus system, the contractor is paid for his actual job costs plus a percentage or a fixed fee for general overhead and profit. In such cases the contract must specify exactly what labor and equipment is to be paid for, the rental rates for the equipment, and what labor and equipment are to be included in the contractor's general overhead. Extensive records must be kept at the job and at the home office covering the cost of materials used, the job payrolls, and the times of use for equipment. These records are kept on a daily basis, certified by the contractor, and checked and approved by the job inspector or by the resident engineer.

On cost-plus work, the owner saves some of the engineering costs otherwise required to prepare a complete set of contract documents and, more importantly in many cases, saves the time required to prepare the complete set of documents and to hold a formal bid opening. In addition, on negotiated contracts, the owner is able to select a particular contractor to do the work, and in some cases business considerations may cause this to be an important feature from the owner's point of view.

On the other hand, cost-plus work normally costs more than competitively bid work. There is less incentive for the contractor to keep costs down, more record keeping is involved, the owner pays for all unforeseen problems, and in many cases, the contractor charges the owner more for his own equipment than he would charge himself if he were bidding the work on a competitive basis.

An additional and serious disadvantage of the cost-plus system is that it does not enable the owner to determine the final cost of the proposed project in advance.

1.6.2 Fixed Price Contracts

Fixed price contracts may be bid as a lump sum for the entire job or on the basis of unit prices applied to the engineer's estimate of the quantities involved for each of the various job items.

1.6.2.1 LUMP SUM CONTRACTS. Buildings are bid on a lump sum basis. In nearly all cases the total quantity of materials and work to be done can be precisely determined from the plans and specifications.

On lump sum bids, the estimator lists the job items on the estimate summary sheet. The listed items correspond to and are in the same sequence as the sections covered by the specifications.

There may be sixty or more sections involved, most of which cover specific subcontract work. The lump sum price bid is the sum of the bare cost of the contractor's own work including his direct job overhead, the subcontractors' prices for their respective sections, and the contractor's allowance for general overhead and profit.

Lump sum contracts require the contractor to construct a complete facility in accordance with the plans and specifications. All required work and materials must be included in the estimate, even though some may not have been covered in the sections of the specifications.

1.6.2.2 UNIT PRICE CONTRACTS. Most engineering work is bid on a unit price basis. On such work it is usually impossible to determine the final quantities for some items from even the best of plans, and the unit price method enables the contractor to receive payment for the actual quantities of work done based upon field measurements made during the course of construction. An important secondary benefit derived from unit price bidding is the relative ease of making preliminary estimates for future work based upon a file of published unit prices bid on similar past jobs.

On unit price bids, the estimator lists the job items and their quantities on the estimate summary sheet. The list is identical to the bid proposal and to the engineer's estimate. The unit prices bid will include the contractor's costs for direct job overhead, general overhead, and profit. Only that work covered by the job items or specified as included in the cost of one or more of the job items need to be done.

The unit *prices* bid for items on the engineer's estimate should not be confused with the unit *costs* applied to the various activities required to complete the items and which are the basis of the cost estimate. In the broadest sense, the unit price bid for any item is determined by dividing its total estimated cost by the number of units on the engineer's estimate for the item.

On most unit price contracts, some of the items themselves are bid on a lump sum basis. Lump sum items save engineering time during the preparation of pay estimates, and the contractor saves time during the bidding process, since it is not necessary to check the engineer's estimated quantities as must be done when the unit price method is used and the pay limits differ from those required to do the work.

Overhead and profit could theoretically be prorated to the bare costs of the various items of work. This is not practical because of lack of time on bid day and uncertainties regarding the final pay quantities of some of the items on the engineer's estimate. The contractor cannot afford to assign appreciable amounts of overhead and profit, which are fixed costs, to an item that may underrun by a substantial amount on completion.

In practice, therefore, overhead and profit are arbitrarily distributed to the major items of the contractor's own work and in particular to items bid as a lump sum, to items whose final pay quantities, as shown on the engineer's estimate, are known to be reasonably correct, and to items which may be expected to overrun the engineer's estimate. Early pay items are favored over late pay items since this enables the contractor to start generating working capital at an earlier date.

1.7 SUBCONTRACTORS

The subcontractor is a specialist in a particular field. He exists because he can do his special work for less money than the general contractor would spend, overall, if he were to do the work himself. The subcontractor furnishes all labor, materials, and equipment necessary to complete a portion of the project. His completed work must conform to the plans and specifications in all respects, but his method of doing his work must be entirely under his own control. He works for and is paid by the general contractor.

Subcontractors make their own takeoffs, quote their work at a fixed price, and furnish their own direct job supervision. As a result, the time required by the general contractor to bid work, his general overhead and job supervision, costs, and the amount of uncertainty at bid time are all reduced. Subcontractors as a group furnish the general contractor with most of the advantages of a greatly increased organization at no additional cost to the contractor or the owner.

Essentially, all subcontract quotations are made by telephone on the bid day and are subsequently confirmed in writing to the low bidder. On building work, subcontractors normally quote a complete section of the special provisions, and in such cases their quotations are noted to be ''plans and specs.'' Special provisions for engineering work cover the work by job items and not by sections relating to subcontract trades. On such work, subcontractors frequently quote only a portion of a particular job item.

1.7.1 The Subcontract Agreement

On being awarded the job, the general contractor enters into subcontract agreements with those subcontractors whose quotations were the lowest at bid time. At the very minimum, the subcontract agreement should clearly specify the work to be done and the method of payment, should bind the subcontractor to the general contractor in the same manner that the general contractor is bound to the owner by the contract documents, and should specify the procedures and time periods to be followed and allowed in case default by the subcontractor requires his removal from the job. On building work, in particular, subcontractors should be required to clean up their own work during the course of the job and upon completion of their work.

1.7.2 Subcontractor's Bonds

The subcontractor's bond is a guarantee that the subcontractor will do the work and pay his labor costs and his suppliers. It is addressed to the general contractor and is signed by the subcontractor and by his bonding company.

The cost of a subcontractor's bond is essentially the same as the cost of the usual contract bonds but is based on the amount of the subcontract. When subcontract bonds are required, their cost is added to the subcontractor's quotation.

1.7.3 Material Suppliers

On building work, a number of major material suppliers do their own takeoffs and quote a fixed price by telephone on bid day. They are similar to subcontractors, but they do not include the installation of their materials. Among such suppliers are those who furnish hollow metal doors and frames, millwork, wooden doors, and finish hardware. In all cases their quotations should be plans and specs for the material furnished.

1.8   JOINT VENTURES

Two or more contractors may join forces and form a partnership or joint venture for purposes of bidding and constructing a particular project. Such a joint venture is a separate legal entity and has its own name, address, and required licenses, and the contract bonds are made in its name.

Joint ventures are sometimes formed when one or more of the partners is unable to bid a proposed project alone because of prequalification or bonding company limitations.

Specifications for engineering work frequently require the general contractor to do 50 percent or more of the work with his own forces. When a proposed project is fairly evenly divided between two primary categories of work, it may be impossible for a contractor specializing in one of them to do 50 percent of the work if he subcontracts the other. In such cases, a joint venture enables two contractors, each an expert in one of the major work categories, to submit a joint venture bid that is lower than would be probable if each were to bid the job alone.

Since a joint venture is a partnership, each partner is responsible to the owner for the successful completion of the project, and it is therefore imperative that each of the joint ventures have complete confidence in the integrity, ability, and financial strength of the others. Should one or more of the joint venturers go bankrupt or fail to perform, the remaining venturers must complete the project regardless of their particular areas of expertise.

Although all joint ventures are treated as separate legal entities by the owner, they may be structured internally either as true joint ventures or as item joint ventures.

1.8.1   True Joint Ventures

In a true joint venture the partners contribute working capital to the joint venture bank account, hire all office and job personnel on the joint venture payroll, buy all supplies, and rent or purchase all equipment in the name of the joint venture. In addition, all invoices and payrolls are paid with checks drawn on the joint venture bank account. On completion of the work the profit made by the joint venture is distributed to the partners as set forth in their joint venture agreement.

1.8.2   Item Joint Ventures

On an item joint venture, each of the venturing partners is responsible to the joint venture for a specific number of job items set forth in the joint venture agreement. The prices each partner quotes to the joint venture for his items include his overhead and profit, and his items normally include those subcontract items relating to his area of work.

One of the venturers acts as the managing partner for the job, and his address is used as the joint venture address. The managing partner combines the prices quoted by the other partners with his prices for his own items and submits the bid.

If the bid is successful, a bank account is set up in the joint venture name for the sole purpose of processing payments from the owner. Such payments are deposited therein and are immediately disbursed to the partners, in accordance with their earnings on their items of work.

On item joint ventures, each partner hires all personnel for his items of work on his payroll and pays all of his suppliers with checks drawn on his bank account. In the broadest sense, the managing partner performs as the general contractor for the job, and the other partners perform as subcontractors.

Upon completion of the work, the amount of profit realized by each venturer will depend on his success in bidding and completing his items of work. On an item joint venture, it is thus possible for one or more of the partners to make more than the estimated amount of profit, while the others may make less.

1.9 PROCEDURE ON BID DAY

The actual procedures followed by different contractors on bid day will vary, but the basic operations required and the conditions that exist are the same for all of them. An appreciation of the conditions confronting the contractor on bid day is essential to the estimator and to those engineers and architects who prepare contract documents and hold bid openings. The hectic situation that exists when large building jobs are bid results from conditions that are beyond the control of the contractor. Bid proposals requiring unessential information or many addenda add to the confusion and result in higher bid prices.

The contractor should have all his own work completed and entered on the summary sheet prior to bid day. The sheets used are of legal size and are taken from standard columnar ruled pads.

1.9.1 Bidding Building Work (Light Construction)

The majority of the subcontractors submit their quotations by telephone during the last 2 hours prior to the bid opening. It is not unusual to receive as many as twenty-five quotations or cuts in prices during the last half hour. The contractor must handle this telephone traffic, enter and revise quotations on the summary sheet, revise the list of subcontractors and keep it in phase with the changing quotations, make a final determination of his allowance for general overhead and profit, determine the total job cost, fill out the bid proposal, and deliver the sealed bid to the designated location prior to the bid deadline. In many instances, the place designated for the bid opening will be several hours away from the contractor's office where the bid is being prepared.

A supply of printed forms on which subcontractors' bids are entered should be available together with a suitable file box for them. In addition to having spaces for the quotation itself, these forms should have a block at the top where the line number of the section quoted, as set up on the estimate summary sheet, may be entered. The quotations are filed numerically by line numbers. Since the contractor must frequently refer to previous quotations, it is imperative that he be able to retrieve them with no unnecessary loss of time.

A typical estimate summary sheet for a reinforced concrete building project is shown in Figures 1.1 and 1.2. The lines are numbered consecutively at the left-hand margin. All costs are entered in column 1, and columns 2 and 3 are used for processing last-minute cuts in prices.

The contractor enters his own name on line 1 together with the total cost for all of his own work; this total includes his cost for direct job overhead.

The estimated cost of the contract bonds is entered on line 2 together with a notation of the total estimated job cost on which the bond cost is based and the rate of change in the bond cost per thousand dollars of bid price.

On the project illustrated, the cost for fire insurance was paid by the owner, and the liability insurances were based on a composite rate and assessed as a part of the payroll expenses, which are discussed in Chapter 3. When applicable, insurance costs normally are entered immediately following the cost of the bonds, although some contractors carry insurance costs on their direct job overhead sheet.

In our example, starting with line 3, the sections from the specifications that are to be subcontracted are listed in the same sequence in which they are covered in the specifications. Those sections relating to the general contractor's own work are omitted since all of his costs were combined and entered on line 1.

A separate sheet on which the names and addresses of the successful (low) subcontractors will be listed is set up in the same order as the estimate summary sheet. This sheet is called the "subcontractors list," and the line numbers for each trade listed are identical to those on the summary sheet. The subcontractor list will not show any of the prices quoted.

By keying the quotation forms and the subcontractor list to the line numbers from the estimate summary sheet, the required entries and their revisions can be made in a minimum amount of time.

From three to six people will be required to take quotations over the telephones. The estimator has the original copy of the summary sheet and the others have duplicate copies. These duplicates are used to key quotations to line numbers, and no prices are entered on them.

Most bid openings are held in the afternoon. At the start of the day, the quotations from subcontractors come in slowly. The quotations are entered on the forms, and the line number for the section quoted is entered in the block provided. Each quotation is given to the estimator, who enters it in pencil on the summary sheet, lists the subcontractor, and files the quotation form numerically by the line number. During the final 2 hours prior to the bid opening, another person will be required to process the subcontractors list, and this person should be within arm's reach of the estimator.

LINE NO.	SEC.		COST	REVISED COST	ADD CUT
1		ABC CO.	1,371,000		
2		BONDS (7500 M, +4.80, -5.25)	48,000	50,000	+2,000
3	2A	EARTHWORK	88,500		
4	2B	A.C. PAVING	3,710		
5	3A	REINF. STEEL	345,000		
6	3B	PRECAST CONC.	69,800		
7	4A	MASONRY	5,910		
8	5A	STRUCTURAL STEEL	117,900		
9	5B	ARCH'L. METAL	44,000		
10	6B	MILLWORK, WOOD DOORS	74,500		
11	7A	MEMBRANE ROOFING	47,200		
12	7B	B'LD'G. INSULATION	16,900		
13	7C	SPRAYED INSULATION	13,500		
14	7D	SEALANTS	3,750		
15	7E	DAMPPROOFING	22,400		
16	7F	AGGRE-DECK	14,500		
17	7G	FLEX-DECK	17,200		
18	8A	STORE FRONT			
19	8B	ALUM. WINDOWS			
20	8C	PORCELAIN PANELS	149,500		
21	8D	ALUM. DOORS			
22	8E	GLASS, GLAZING			
23	8F	AUTOMATIC DOORS	21,000		
24	8G	H. M. DOORS, FRAMES	49,100	44,000	-5,100
25	8H	FOLDING DOORS	10,500		
26	8J	FINISH HARDWARE	58,500	57,500	-1,000
27	8K-2	VAULT DOOR	1,250		
28	8K-3	ROLLUP DOORS	1,650		
29	8K-4	FIRE DOORS	5,200		
30	9A	LATH & PLASTER	532,500	491,000	-41,500
31	9E	ACCOUSTICAL	28,500		
32	9F	PAINTING	70,000		
33	9G	CERAMIC TILE	59,500		
34	9H	TERRAZZO	49,400	52,000	+2,600
35	10A	TACK, CHALK BOARD	430		
36	10B	TOILET COMPARTM'TS.	2,300		
37	10C	LINEN CHUTES	6,100		
38	10D	METAL LOCKERS	6,950		
39	10E	TOILET ACCESSORIES	26,700		
40	10F	CUBICLE TRACKS	12,600		
41	10G	NAME PLATES	2,600		

SUMMARY - CITY HOSPITAL

Figure 1.1. Estimate summary sheet.

As subsequent quotations are received, the estimator compares them with those previously entered on the summary sheet, erasing the previous entries and inserting the new ones if they are lower. The subcontractor list is revised to correspond, and the quotation forms filed by line number, with the low quotation in front.

Forty-five minutes prior to bid opening, all telephones will be ringing constantly,

SUMMARY – CITY HOSPITAL (CONT'D.)

LINE NO.	SEC.		COST	REVISED COST	ADD CUT
42	10H	INCINERATOR	3,920		
43	11A	KITCHEN EQUIPM'T.	69,900		
44	11B	MODULAR MET. CASEWORK	253,000		
45	11C	AUTOPSY EQUIPM'T.	10,500		
46	11D	JANITOR STATION	7,350		
47	11E	CHART·HOLDER	2,890		
48	11F	FILM ILLUMINAT'N.	4,560		
49	11G	KITCHEN UNIT	700		
50	12A	DRAPERY TRACK	3,450		
51	12B	CARPET, RESILIENT FLOOR	103,900	101,000	−2,900
52	12C	VERTICAL BLINDS	8,400		
53	13A	RADIATION PROTECT'N.	11,800		
54	13B	MOVABLE WALL SYST.	112,500		
55	14A	PASS. ELEVATORS	211,200		
56	14B	FREIGHT ELEV. (EST)	30,000	27,000	−3,000
57	15A	SHEET METAL			
58	15B	HEAT'G. & VENTILAT'G.			
59	15C	PLUMBING	2,319,600		
60	15D	MEDICAL GASES			
61	15E	FIRE SPRINKLERS	28,500		
62	16A	ELECTRICAL	1,061,700	1,020,000	−41,700
63			—	CUT	−90,600
64			—		
65		TOTAL LESS (3)	5,142,820		
66		TOTAL (3)	2,499,100		
67		TOTAL INC'L. (3)	7,641,920		
68		G.O. + PROFIT	350,000		
69		TOTAL	7,991,920		
70		CUT	90,920		
71		BID PRICE	7,901,000		

Figure 1.2. Estimate summary sheet (continued).

and there will normally be three or four sections for which no quotations have been received. The estimator encircles these open spaces in the cost column to make them easily identifiable and makes a duplicate copy of his summary sheet. In our example these openings are centered on lines 20, 56, and 59. At the time the duplicate copy is made, most of the quotations listed in column 1 will have been revised from one to four or more times. The total for column 1 less the encircled openings is entered on the estimator's duplicate summary sheet; in our example this total is shown on line 65. This procedure enables most of the addition to be made and checked prior to the last-minute rush.

From this point on, the estimator works with the duplicate summary sheet, and there is literally no time to spare. No further erasures are made in column 1, and subsequent changes in prices are processed in columns 2 and 3. In our example, last-minute revisions are shown on lines 2, 24, 26, 30, 34, 51, 56, and 62.

During this time, the estimator must normally contact one or more subcontractors for clarification of their quotations. In addition, combined quotations for several sections may be received and must be compared with the separate quotations for the same sections. This comparison is best made on a separate worksheet, as discussed in what follows.

When the designated location for the bid opening requires more than 10 minutes of travel from the contractor's office, a satellite employee with the signed bid proposal and subcontractor list is positioned at a telephone near the designated location. In those cases where the satellite employee is located several hours away from the home office, he will be in constant communication with the office during the last half hour. He is given the list of subcontractors as it then stands, makes the listing in pencil, and revises it as the subcontractors change. Since the list is keyed to the line numbers on the estimate summary sheet, this procedure is fast and there is little chance for error.

Twenty minutes prior to bid time, all openings in column 1 of the summary sheet will normally be filled, but all telephones will be ringing, as the subcontractors cut their earlier quotations. In those cases where an encircled opening has not yet been quoted, the estimator inserts an estimated value for the item and determines the sum of all of the encircled quantities. Thus, in our example, an estimated price of $30,000 is entered in the opening at line 56, and this figure is corrected, when the quote is received, in columns 2 and 3.

The total for the encircled quotations is entered (line 66 in our example) and added to the previous total. The contractor's final allowance for general overhead and profit (discussed in Chapter 3) is finalized and added to the previous total to arrive at a bid price which is approximate since the cuts in column 3 of the summary sheet have not been made.

The cost for the contract bonds is revised to correspond to the approximate bid price, and the change in cost is entered in column 3 of the summary sheet. In the example this operation is shown on line 2.

The additions and cuts shown in column 3 are added algebraically, and the resulting cut is deducted from the approximate bid price to obtain the final bid price for the job. In the example this process appears on lines 70 and 71. In a typical case, the cut will be made about 10 minutes prior to bid time, and the telephone traffic will have decreased. When additional last-minute cuts are received, they are processed in columns 2 and 3 of the summary sheet using a colored pencil for easy identification. The bid price is revised to correspond and is given to the satellite employee, who enters the bid price, in ink, on the proposal, makes the final revision to his subcontractor list, seals the proposal in a preaddressed envelope, and delivers it to the designated location prior to the designated time for the bid opening.

Alternates, when required, are usually processed on separate summary sheets for each alternate. On these sheets, only those sections relating to the alternates are listed. Simple alternates involving only two or three trades are sometimes processed on the estimate summary sheet, which will require three additional columns for each alternate.

1.9.1.1 PROCESSING COMBINED QUOTATIONS. During the last hectic hour, the contractor is frequently confronted with combined quotations for two or more sections which are also being quoted separately by other subcontractors. To avoid confusion and error, he must be prepared to evaluate such combinations in a systematic and orderly manner, and the procedure should be such that the comparison can be made by a qualified assistant.

The comparisons are made on a separate sheet, such as the one shown in Figure 1.3, and the format for this comparison sheet should be set up prior to the day of the bid opening. Usually three job sections, or trades, will be involved.

On the comparison sheet shown in Figure 1.3, the trades quoted separately and in combinations are listed on line 1 together with their respective line numbers from the estimate summary sheet. Line 2 shows the separate quotations for the three sections as entered on the estimate summary sheet. The total for the three sections is listed in column E. The names of the subcontractors submitting these quotations are entered on the subcontractor list at the time the quotations are entered on the summary sheet. After the quotations have been entered on line 2 of the comparison sheet, no changes, erasures, or comparisons of these sections are made on the summary sheet, and no changes are made in line 2 of the comparison sheet.

		A	B	C	D	E
	COMBINED QUOTATIONS					
1	SECTION (LINE NO.)	DRYWALL (32)	ACCOUSTIC. (34)	LATH & PL. (39)	TOTAL	TOTAL
2	ENTERED IN COST COLUMN	53,000	31,000	310,000		394,000
	SUBCONT'R.	LOW QUOTATIONS				
3	JONES	52,000				
4	BROWN		30,000			
5	SMITH			300,000	382,000	
6	AJAX	80,000	INCL.	—	380,000	
7	KELLEY	—	INCL.	327,000	379,000	
8	ACME	INCL.	INCL.	INCL.	375,000	375,000

CUT 19,000

Figure 1.3. *Comparison sheet for combined quotations.*

Comparison sheet lines 3, 4, and 5 show the low quotations and the subcontractors' names for the separate quotations subsequently received. These data may be revised as necessary. The total for the three separate low quotations is entered on line 5, column D.

Lines 6, 7, and 8 show the names and the quotations for the low subcontractors submitting combined quotations. Thus, Ajax quotes $80,000 for drywall plus accoustical, which is added to the $300,000 shown on line 5 to arrive at a total for the three sections, which is entered on line 6, column D.

Similarly, Kelley, line 7, quotes $327,000 for lath and plaster plus acoustical, which is added to the $52,000 shown on line 3 to arrive at a total for the three sections, which is entered on line 7, column D.

Acme quotes all three sections for a total of $375,000, which is entered on line 8, column D. The data entered on lines 6, 7, and 8 may be revised as necessary when lower quotations are received.

The totals shown in column D, lines 5 through 8, represent the low combinations for the three sections, and the lowest one, on line 8, is entered in column E. The $19,000 cut is the difference between this figure and the $394,000 total for the three sections as entered on the estimate summary sheet. At the time this cut is transferred to the estimate summary sheet, the subcontractor list is revised to show Acme as the subcontractor for lines 32, 34, and 39.

1.9.2 Bidding Engineering Work (Heavy Construction)

On engineering work, fewer subcontractors are involved, the telephone traffic is relatively light, no subcontractor list is required, and there are normally no alternates. On the other hand, the estimate summary sheet and the bid proposal may contain a hundred or more job items and consequently require more time to complete.

The bid proposal form contains a consecutively numbered list of the job items and their estimated quantities. Spaces are provided for listing the unit price and the total price bid for each job item. The job price is the sum of the prices bid for the listed job items.

The estimate summary sheet is similar to the bid proposal except that it has an additional column for listing the bare cost of each item. The contractor's direct job overhead cost and his allowance for general overhead and profit are listed in this cost column below the last job item. Since the proposal does not include items for overhead or profit, these costs must be distributed among and included in the prices bid for the various job items. On completion, the total for the cost column will equal the total for the total price bid column.

Typical estimate summary sheets for small projects are shown in Figures 8.10. Sheets from standard $8\frac{1}{2} \times 14$ four-column pads are normally used. The list of job items is double spaced and is identical to the one on the bid proposal. It is imperative that the quantities listed on the summary be checked against those shown on the proposal since all extensions are made and checked on the summary prior to being entered on the proposal itself.

On the day before the bid opening, the contractor lists the bare cost for each of his items and his direct job overhead cost on the summary sheet. Some items will be partially subcontracted, and the subcontract costs, when received, will be entered in the double space below the item involved. Horizontal lines are drawn through all unneeded double spaces in the cost column and through all double spaces in the total price bid column to ensure that all required spaces in both columns will be visible and ultimately filled. On completion, these two columns will have a number or a line in every space.

The unit prices for all but the major items of the contractor's own work are determined by dividing their bare costs by the number of units involved. The results are rounded off, entered, and extended on the summary sheet.

Estimated unit prices for minor subcontract items together with their extensions are entered on the summary sheet even though the actual quotations will be received or entered in the cost column until the following day. With the summary sheet thus filled out as far as possible, the available prices are entered on the bid proposal in ink. As each item is entered on the proposal, a hashmark is drawn through the checkmark on the corresponding total price on the summary sheet. This enables the contractor to determine which items have entered on the proposal by scanning the summary sheet and is important in those cases where the signed proposal will be in the hands of a satellite employee on bid day.

On bid day, the subcontractors will quote on a unit price basis. These unit prices are multiplied by the quantities involved, and the resulting costs are entered in the cost column of the summary sheet. As a general rule, the unit prices quoted are rounded off and entered, together with their extensions, in the price columns of the summary sheet. Where large subcontracts are involved, the contractor may assign a portion of his costs for overhead and profit of the item.

The costs for the contract bonds and special insurance are normally carried on the direct job overhead sheet on unit price bids. Revisions to these costs and last-minute changes in subcontractors' quotations are processed on a separate cut sheet, as discussed in the following.

Ninety minutes before the bid opening on larger jobs, there may be four or five subcontract items that have not been quoted and will be open in the cost column of the summary sheet. These same items together with two or three of the contractor's own major items will also be open in the total price column. The estimator encircles the open items in each column to ensure their easy identification and enters the totals for each column, less it encircled openings, on the summary sheet. The total for the cost column includes the cost of the contractor's direct job overhead. The preceding procedure enables the contractor to complete and check the bulk of the required addition prior to the last-minute rush.

In cases where a satellite employee has the bid proposal at a distant location, he will, by this time, have the proposal filled in except for the open items, and he will keep the bid proposal current from this time on.

Forty minutes prior to bid time, all open spaces in the cost column will be filled. Their sum, plus the cut from the cut sheet, are added to the preceding total to obtain the bare cost of the job at that time. The allowance for general overhead and profit is finalized and added to the bare cost to obtain a preliminary total cost, which will be revised as later cuts are received.

The remaining unit prices are established as discussed in Section 1.6.2.2 but with the added criterion that the total price bid must equal the total job cost. The unit prices for all but one of the open items are entered on the summary sheet and the proposal together with their extensions. The sum of these prices is added to the previous total for the total price column to obtain the total bid price for the job less the bid price for one open item. During this period, additional cuts are being processed on the cut sheet using a colored pencil, and the cuts are listed in a second column designated "cut number 2."

Fifteen minutes before bid time, cut number 2 is totaled, and the final job cost is determined. The difference between this cost and the last total for the price column represents the total price for the open item, which will exactly balance the bid price with the total job cost. In practice, the resulting unit price for the open item is rounded off, and the total price for the item, based on the rounded-off unit price, is added to the previous total to obtain the total bid price for the job. The prices for the open item and the total job price are entered on the bid proposal, which is sealed and delivered to the designated location prior to the designated time for the bid opening.

1.9.2.1 THE CUT SHEET. Unit price jobs require an estimate summary sheet which is considerably more complicated than the ones used for lump sum bids. To avoid confusion, late revisions to subcontractors' quotations, material prices, and certain direct job overhead items are processed on a separate cut sheet.

A typical cut sheet is illustrated in Figure 1.4. Those items that require price revisions are listed in column 2 together with their quantities. Space is allowed to enter the subcontractors' quotations below each item. The item numbers are entered in column 1; the bond cost, a direct job overhead item which has no number, is designated as DO.

The cost of each item, as entered on the estimate summary sheet or on the estimate sheets for the items, is listed in column 3. After these costs are entered in column 3, all revisions will be processed on the cut sheet.

A bond cost of $10,100 was entered on the direct job overhead sheet. This cost was based on an estimated total job price of $1,500,000. The rate of change in the bond cost is $5.60 per thousand dollars of total job price, as noted in column 2. The final job price is $1,410,000, and this decrease in price results in the $500 cut shown in column 5.

In estimating the cost of item 8, an estimated price of $35 per cubic yard was used for the concrete cost since no firm quotation was available at that time. A subsequent quotation of $33 per cubic yard results in the $1200 cut shown in column 5.

Item 9 indicates that an estimated cost of $25 per CWT (hundredweight) was used on the estimate summary sheet. A subsequent quotation of $26 per CWT results in an increased cost of $1500, which is listed in column 5.

Item 54 indicates a typical cut resulting from a lower quotation.

1	2	3	4	5	6
NO.	ITEM AND UNIT COST	ENTERED	REVISED	CUT NO. 1	CUT NO. 2
DO	BONDS EST. 1500 M 5.60 $/M FINAL 1410 M = −90 M	10,100	9,600	− 500	
8	CONCRETE 624 CY EST. 35 CONCRETE INC. 33	21,840	20,590	−1200	
9	REINF. STEEL 1500 CWT EST. 25.00 JONES 26.00	37,500	39,000	+1500	
54	PRECAST GIR'S. 10 EA UNITED 2600 MORGAN 2300	26,000	23,000	−3000	
				−3200	

Title: CUT SHEET

Figure 1.4. *Cut sheet.*

Cut number 1, as shown, equals $3200. Later changes in cost are processed using a colored pencil, and the resulting cuts are listed in column 6.

1.10 EQUIPMENT COSTS

In preparing cost estimates, the cost for heavy equipment is based on an hourly rental rate in the majority of cases. The rental rate used will include the costs of operation, maintenance, insurance, depreciation, and subsistence for the operators when required.

Most contractors do not have a daily or continuing need for heavy equipment, and it is rented when needed from equipment rental firms or from other contractors at prevailing rental rates. Because of competitive pressure in the equipment rental business, the established rental rates are reasonably close to the true cost of owning, operating, and maintaining the particular piece of equipment.

Earthwork contractors frequently own a basic string of earth-moving equipment. In such cases, estimates involving considerable use of the equipment may be made using rental rates based on the contractor's actual hourly costs for owning, operating, and maintaining the equipment. The rates are determined from past job cost records, and in those cases where the contractor conducts an efficient operation, they will be less than the prevailing rental rates.

Prevailing rental rates are always used in estimating job costs when contractor-owned equipment will only be required for short period of time. In such cases the contractor's equipment may be working on another job when needed, or good operators may not be available on a short-term basis.

When a contractor rents the heavy equipment as needed, less capital is required and he is not faced with the need to keep the equipment working throughout the year. In the estimates that follow, equipment costs are based on assumed prevailing rental rates.

1.11 ESTIMATING TIME REQUIRED FOR COMPLETION

At the time a job is being bid, it is seldom necessary to prepare a formal critical path diagram in order to estimate the time required to complete the project. Those job items on the critical path for the project are normally apparent to the estimator by the time the cost estimates for each item have been completed, and only those items on the critical path are considered. After the bid opening, the successful contractor prepares a time schedule for the job based on a critical path diagram.

The estimate of the time required to complete each job item is made on the estimate sheet for the item after the cost to perform the work has been determined. The simple procedures required to prepare time estimates are illustrated in detail in each of the estimates that follow.

When labor costs represent the major cost element for a given operation, the time required to complete the operation is determined by dividing the total labor cost for the operation by the daily cost of the proposed crew. When equipment costs represent the major cost element, the time required is found by dividing the total number of units to be handled by the daily production rate of the proposed string of equipment.

The working days so determined for each operation are calculated and listed in the "days" column of the estimate sheet, and if necessary, some of these times may be reduced by increasing the number or the size of the proposed crews or strings of equipment. The sum of the listed times represents the number of working days required if each operation were performed consecutively. The listed times are then modified to conform to the estimated job scheduling which will entail doing as much of the work concurrently as is practicable. The sum of these adjusted times represents the number of working days required to complete the item.

In a similar manner, the number of working days required for each item on the critical path for the job is tabulated together with an estimated time for mobilization and moving off on completion of the work. Included on this list are the estimated times required to complete certain subcontract items; in some cases conversation with one or more subcontractors may be required to determine a realistic time for their work. The sum of the times listed for the various items represents the number of working days that would be required to complete the job if each item were done consecutively. As before, the times required for each item based on overlapping the work as much as may be practical are listed in a second column, and their sum represents the number of working days required to complete the project.

Additional working days are added to allow for the time that will be lost in winter because of inclement weather. The size of this allowance will depend on the geographical area involved; for example, about 20 working days are allowed, per winter, for work performed in Southern California. Assuming 6 holidays per year, the total number of working days is divided by 21.1 to convert the time required for completion to calendar months.

Holidays affect the calendar time required for completion but do not affect unit costs or production rates. Bad weather increases both unit costs and the time required for completion. The effect of weather conditions on the unit costs used will depend on the circumstances and on the estimator's judgment. Unit costs and production rates from past jobs reflect the effects of the weather on the jobs considered. Disastrous results have been realized, however, when unit costs based on California conditions were indiscriminately used to bid work in Alaska.

The estimated time for completion determines an important component of the total cost for any project since direct job overhead costs depend upon the duration of the job. The time required for completion can be quickly determined when the estimate sheets for the various job items have been properly set up, but the estimator must exercise care and judgment in the process and must not overlook the importance of the time estimate to the determination of total job cost.

Specifications will indicate the length of the project if ''liquidated damages'' are called for.

1.12 THE COST SYSTEM FOR WORK DONE WITH OWN FORCES

A cost system based on account numbers for each job item is essential if unit costs are to be developed and if the work in progress is to be properly monitored and controlled. To prevent serious errors of omission, the cost system should be an integral part of the contractor's bookkeeping system. It should furnish the contractor the information he requires and should be standardized as much as possible.

Contractors are inherently highly independent individuals. Their cost-keeping systems are essential to their survival and, in most cases, have been developed by them over the years to fill their specific needs. In many cases, the contractor's cost system and his unit costs are closely guarded. Beginning contractors and young engineers starting a file of unit costs based on their own job experience will, however, need a cost system, and the one described next will be more than adequate.

A list of account numbers is kept at the home office and at each job, and the time for each workman and for each piece of equipment is distributed to the proper accounts in the field on time cards. The distribution may be made by the foremen using one time card for each crew of workmen. The time cards are extended to

ABC Company

JOB NO. 64 Date _____

Account no.	Sub acc't. no.	Description	A. Jones	B. Smith	C. Woods	D. Brown	E. Green	A. Wilson	B. Thomas	ABC #8	Ajax Crane	01 Labor	03 Rental
Emp. No.													
Class.			L	L	C	C	C	C	E				
Rate ST / OT			9.00	9.00	12.75	12.00	12.00	12.00	13.00	25	60	Labor	Rental
26	48	Conc. pipe – lay	4	4							4	72.00	240.00
09		Str. ex.								8			200.00
0705	0301	SS wingwalls	4	8	8	8	8					462.00	
01	0801	Backhoe						8				104.00	
		Totals	8	8	8	8	8	8	8	8	4	638.00	440.00

Employee hours | Eqpt. hrs | Dollars

Figure 1.5. Time card.

dollar amounts in the office and are posted in the cost ledger for the job. All delivery tickets for materials and rented equipment are approved at the job level, and the proper account numbers are written on them at that time.

An account numbering system using two digits to denote each element of the account offers great flexibility and can be adapted to all situations. Examples of account numbers that might be used are given in what follows, and the time card illustrated in Figure 1.5 is made out using these numbers as an example. In all cases the last two digits of the account numbers will be 01, 02, 03, or 04 and will denote that the cost involved is for labor, material, rental, or subcontract expense.

The account number 64264801 is assumed to read as follows:

	64	job number
	26	rubber gasketed concrete pipe
	48	48 inches in diameter
	01	labor
or	02	material
or	03	rental
or	04	subcontract

The account number 64090001 is assumed to read as follows:

	64	job number
	09	structure excavation
	00	an open subaccount if needed
	01	labor
or	02	material
or	03	rental
or	04	equipment

In the preceding examples 26 and 09 are account numbers and 48 and 00 represent subaccounts. Subaccounts may be further broken down when necessary. Thus the account number 640705030101 is assumed to read as follows:

64	job number
07	bridge concrete
05	bridge number 5
03	abutments and wingwalls
01	set and strip
01	labor

In the preceding example the list of account numbers for subaccount 03 might be shown as follows:

03	abutments and wingwalls
01	set and strip

    07   fine grade
    08   pour

The 01, 07, and 08 numbers would be standard for all concrete items and would be quickly understood and memorized by all of the personnel involved.

The account number 6401080101 is assumed to read as follows:

    64   job number
    01   contractor owned equipment
    08   equipment number 8
    01   operating expenses
    01   labor

It should be noted that B. Thomas, the operator of the company-owned backhoe (ABC #8), is charged to the equipment account but the rental cost for the equipment operated and maintained is charged to the structure excavation account. This enables job unit costs to be developed for the equipment indexed to the rental rate charged for the backhoe and the cost of owning and operating the equipment to be developed in the equipment account. In addition, by crediting the equipment account for the rental charged to the jobs, the equipment account will indicate whether or not the contractor is making money on the equipment.

General Job Costs

General job costs are those costs incurred from running the job as a whole. These costs are not directly related to the performance of any particular item of work. They consist of direct job overhead, payroll expenses, the job's share of the contractor's general overhead, and the job profit.

2.1 DIRECT JOB OVERHEAD (FIELD COSTS)

Direct job overhead is the direct cost of running the job. Before completing the direct job overhead sheet, the estimator must have an estimate of the time required to complete the work, a close approximation of the total job cost, and familiarity with all of the job items. As soon as the estimator has a good feeling of the project, a cost sheet should be prepared.

2.1.1 Checklist for Direct Job Overhead

Direct job overhead includes the costs of those items on the following list that apply to the particular job being bid.

1. Contract bonds.
2. Insurance.
3. Permits, fees, and tests.
4. Supervision and job office salaries.
5. Clearing and grubbing.
6. Office and sheds.
7. Power.

8. Water.
9. Telephone.
10. Radios.
11. Sanitary facilities.
12. Drinking water.
13. Vehicles.
14. Hoisting.
15. Job signs.
16. Layout.
17. Watchmen.
18. Access roads.
19. Fences.
20. Heating.
21. Medical and safety.
22. Punch list.
23. Cleanup.
24. Traffic control.
25. Wage raises.
26. Subsistence.
27. Special requirements.

2.1.1.1 CONTRACT BONDS. Contract bonds were discussed in Section 1.4.4. In every case, the costs for the bonds are obtained from the contractor's bonding company since both the rates and the work classifications to which they apply are subject to change.

On lump sum jobs, as illustrated in Figure 1.1, the bond cost is usually the second item on the estimate summary sheet. On unit price jobs, the bond cost is usually the first item on the direct job overhead sheet, as illustrated in Figure 8.9.

At the time the estimate sheets are set up, the cost for the bonds is based on an approximate job price, and the bond cost is revised, if necessary, when the final job price is determined. The bond costs shown on the estimates presented in Chapter 8 are based on an assumed rate of $7.50 per thousand dollars of total job price.

2.1.1.2 INSURANCE. The costs for insurance represent a significant portion of the overall job cost. For bookkeeping purposes, all insurance costs, with the exception of the cost for the contractor's equipment floater insurance, are considered as direct job overhead items.

The costs for workmen's compensation insurance, public liability and property damage insurance, and in some cases, all liability insurances are based on the contractor's payroll. For estimating purposes, the costs for these payroll insurances are determined and entered on the estimate sheets for each job item as explained in the following section on payroll expenses.

Those insurance costs not based on the payroll are carried on the estimate summary sheet, on building work, and on the direct job overhead sheet on unit price work. In both cases the charges entered initially are based on estimated job costs and are revised after the final job cost is determined.

Workmen's Compensation Insurance.   The cost for this insurance is always based on the job payroll, as discussed in the section on payroll expenses.

Public Liability and Property Damage Insurance (Exclusive of Automobile Insurance).   The costs for this insurance are also based on the job payroll, as discussed in the section on payroll expenses.

Contingent Liability and Property Damage Insurance.   This insurance indemnifies the contractor in the event of claims for injuries or damages resulting from the activities of subcontractors or other parties, for which the contractor may be held responsible. The cost for this insurance may be expressed as a percentage of the total price of the subcontracted work or it may be included in a composite rate for all liability insurances, which is applied to the job payroll.

When the cost is based on a percentage of the total price of subcontract work, the percentage rate used is normally the same for all subcontracts on a given job.

Products and Completed Operations Liability Insurance.   This insurance indemnifies the contractor in the event of claims for injury or damages to third parties arising from the normal use of the completed project. The cost for this insurance may be expressed as a percentage of the total job price or it may be included in a composite rate for all liability insurances, which is applied to the job payroll.

Public Liability and Property Damage Insurance on Automobiles and Pickup Trucks.   The cost for this insurance may be a stated amount per vehicle per year or it may be included in a composite rate for all liability insurances, which is applied to the job payroll.

When the cost is on a per vehicle basis, the cost may be included in the rental rate charged for the vehicle on the direct job overhead sheet.

Excess Limits Insurance or Umbrella Insurance.   Either of these insurances provide increased limits for the primary liability insurances. An umbrella policy will, in addition, provide certain additional coverages which are usually subject to a deductible of $10,000 or more. The costs for the umbrella policy have become excessive, and the excess limits insurance is often used. The costs for these insurances are a percentage of the total premium charged for all of the primary liability insurances subject to a minimum premium, but the cost may be included in a composite rate used for all liability insurances, which is assessed to the job payroll.

Fire Insurance.   Fire insurance is always required on building work, although its cost is sometimes paid by the owner. The cost will be a specified percentage of the bid price for that portion of the work which is subject to fire damage.

Installation Floater Insurance. This insurance indemnifies the contractor in the event of job accidents resulting in damages to portions of the completed work. The insurance may be purchased for protection against a particular hazard or it may cover all risks. The price is quoted by the insurance agent for each specific job.

The cost for installation floater insurance is often prohibitive. In such cases, the cost for those additional steps necessary to substantially reduce the risks from particular hazards is included in the costs for the job item involved.

Contractor's Equipment Floater Insurance. This insurance indemnifies the contractor should his equipment be damaged by accident or vandalism. In most cases, only motorized equipment is insured, and the insurance costs are included in the rental rates charged for the equipment.

2.1.1.3 CLEARING AND GRUBBING. The cost of incidental clearing and grubbing is included here in those cases where no job item was set up for this work.

2.1.1.4 OFFICES AND SHEDS. This item covers the costs of setting up and removing the job office, sheds, and yard.

2.1.1.5 POWER. This item covers monthly power bills. The costs to connect and disconnect the power source are included in the cost of the job item requiring power.

2.1.1.6 WATER. This item covers the monthly water bills for incidental amounts of water. The costs of developing the water supply and applying such water are included in the cost of the item requiring the water. The costs relating to large amounts of water, like those required for earthwork compaction, are included in the cost of the job item involved or in special job items for these costs.

2.1.1.7 RADIOS. This item covers the monthly rental charges for the mobile radios used.

2.1.1.8 SANITARY FACILITIES. This item covers the cost of the job toilets and other sanitary facilities as required. The cost is based on monthly rental rates which include servicing and moving the facilities when necessary.

2.1.1.9 DRINKING WATER. This item covers the cost of water cans, ice, and paper cups used to supply drinking water to the crews.

2.1.1.10 VEHICLES. This item covers the monthly rental charged for pickup trucks, flat rack trucks, and the superintendent's car. The rates used include fuel, maintenance, repairs, and depreciation.

2.1.1.11 HOISTING. This item covers the costs of hoisting men and materials. It is primarily of importance on building work, where the specifications frequently require the general contractor to furnish hoisting facilities for all of the trades.

On buildings up to three stories in height, personnel elevators are not required and materials hoisting may be done with truck cranes. Although the general contractor will have included the costs for hoisting in his cost estimates for his own work, the requirements of each subcontractor must be considered and an allowance made for the additional crane rental involved.

On tall buildings one or more hoisting elevators are required for the workmen and for materials. On larger work tower cranes may be used to handle materials. In both cases the equipment is usually set up initially by the supplier for a fixed fee and rented to the general contractor on a monthly basis.

In all cases the operators are paid by the contractor, who also adds to the height of the elevator and jumps the tower crane up with his own forces as the building rises. On completion of the work the contractor usually takes down the equipment and returns it to the supplier using his own crews for the dismantling operation.

2.1.1.12 LAYOUT. This item covers the cost of job layout when it is not furnished by the owner. Incidental survey work by job personnel is not considered.

2.1.1.13 FENCES. This item covers the cost of temporary fences required to protect the job site. Permanent fencing is carried as a separate job item and is normally furnished and installed by a subcontractor.

2.1.1.14 HEATING. This item covers the cost of equipment and materials required to protect the work from cold weather damage.

2.1.1.15 PUNCH LIST. The punch list is the list of minor repairs and corrections that must be completed prior to final acceptance of the work by the owner. This item is only significant on building work and will depend on the quality of the job supervision and inspection. On office buildings and schools, where $K_{6+2} = 1000$, the cost may run about \$35 per MSF thousand square feet of floor area for the general contractor's own work.* Each subcontractor completes his own punch list items at his own expense.

2.1.1.16 CLEANUP ON ENGINEERING WORK. On engineering work this item covers the cost of the final cleanup required to obtain acceptance of the work by the owner. The work consists primarily of dressing up the job area with a blade, and the costs of such work are discussed in the chapters on earthwork. No additional charges for day-to-day cleanup are required on engineering work since these costs are included in the unit costs developed for the various items of work.

2.1.1.17 CLEANUP ON BUILDING WORK. The cost of cleanup is a significant item on building work; it includes the cost of day-to-day general cleanup in addition to the cost of the final cleanup required by the specifications.

On building work there are many subcontractors, and the work is done indoors. Even though each subcontractor is required to clean up his own work on a continu-

*See Section 3.8 for an illustration of K_{6+2}, the labor index for eight workers.

ing basis, there is an inevitable accumulation of dirt and debris that must be removed by the general contractor in the interests of good housekeeping and overall job efficiency.

The final cleanup required on building work is always covered in the specifications and normally includes the following items:

1. Dressing up the exterior site. This work is similar to that required on engineering work.
2. Sweeping all floors, dusting cabinets, doors, and electrical fixtures, and cleaning toilet fixtures.
3. Washing windows and glass doors and waxing and polishing some floors. Such work is normally done on a subcontract basis by a janitorial firm.

In view of the variables involved, it is not possible to estimate the cost of cleanup on building work with much precision. It is probable that the costs actually realized by the same contractor on two similar building jobs having different superintendents and subcontractors would vary by as much as 20 percent. As an order of magnitude, the labor cost of the day-to-day cleanup plus the sweeping, dusting, and cleaning noted in item 2 has run about $190 per MSF of floor area on buildings ranging from 80 to 200 MSF in size, where $K_L = 13.48$ (K_L is the labor index for one laborer).

The proposed cost for cleanup labor should be considered with respect to both the floor area and the number of laborer hours represented per day of job duration. The final cost selected will be based on the estimator's judgment augmented by cost records from similar work.

Since burning is now prohibited in most areas, an additional allowance must be made for removing the debris from the job site. In most cities there are firms that supply debris boxes and dispose of their contents as required on a monthly rental basis.

2.1.1.18 TRAFFIC CONTROL. This item covers the cost of the flares, barricades, traffic control signs, and flagmen needed to protect the public during the course of the job. The flares, barricades, and signs are normally rented, and the rental charge includes maintenance and replacement as required. When possible, traffic control costs should be included with the cost of the primary job item requiring such control.

2.1.1.19 WAGE RAISES. This item covers the additional costs for labor and equipment resulting from wage raises that occur during the course of the job. In such cases, a portion of the work will be done under the existing rates, and the remainder will be done under the new higher rates. Wage raises are listed on the direct job overhead checklist as a reminder. The additional costs resulting from wage raises are estimated on the estimate sheets for each job item.

The date on which the wage raise will occur must be located on the time schedule for the job. As discussed in Section 1.11, the estimated time required for completion and the job time schedule can only be determined after the cost estimates for each job item have been completed.

When the wage raise occurs during later stages of the job, costs for all of the job items are estimated on the basis of existing rates, a job time schedule is set up, and the point on the schedule where the wage raise will occur is determined.

The labor and equipment costs for each job item or portion of a job item that will be performed after the wage raise are then increased by an appropriate percentage. The percentage used is determined from the ratio of the cost indexes (Section 3.8) applicable after and before the wage raise. Thus if $K_{6+2} = 1000$ based on the existing rates and if $K_{6+2} = 1075$ after the wage raise, then those labor costs for carpentry work occurring after the raise would be increased by 7.5 percent.

When most of the job is done after the wage raise is in effect, the estimated cost of all of the job items is based on the increased rates, and deductions are made on the estimate sheets for the work that will be performed under the existing rates.

The effect of wage raises on rental rates depends on decisions made by the owner of the equipment. A raise of $1.50 per hour for an equipment operator will often result in an increase of $3.00 per hour in the rental rate charged.

2.1.1.20   SUBSISTENCE. Some contractors convert the specified daily payments for subsistence, when required, into hourly rates that are treated as additional wages. In all cases, the applicable hourly rates are added to the rental rates charged for operated equipment.

The requirements for paying subsistence are somewhat complex, and it is not possible to convert a specified daily rate into an exact hourly rate; therefore, the following procedure is recommended for estimating the costs of subsistence payments for job labor.

The total labor cost (exclusive of payroll expenses, Section 2.2) for each job item is increased by an appropriate percentage to cover the estimated costs for subsistence. The composition of a typical crew for each job item is assumed, and the percentage used is based on the ratio of the estimated daily costs for the crew with and without subsistence. The allowances made for subsistence on labor are then collected and monitored as a direct job overhead item, for control purposes, on those jobs awarded to the contractor.

This procedure results in unit costs for labor that are not affected by variations in subsistence payments and that can be interpreted on their own merits. Similarly, the allowance made for subsistence can be monitored on its merits and adjusted on future estimates, if warranted.

2.1.1.21   SPECIAL REQUIREMENTS. The preceding contain those direct job overhead items that occur regularly in practice. This item includes the costs of additional direct job overhead items that may be required by a particular set of plans and specifications.

2.2 PAYROLL EXPENSES

Payroll expenses are the costs of certain taxes and insurances which are based on the contractor's total payroll costs for all jobs in progress. These expenses are allocated to each job in proportion to its payrolls and are monitored against the allowances made on the estimate for control purposes. The cost of the payroll expenses is estimated by applying a single composite percentage rate to the total labor cost for each job item.

The true cost of each payroll expense item is determined by applying a specified percentage rate to a dollar amount which is based on the employee's base wage rate plus the hourly rate for the employee's vacation plan. For purposes of this discussion, dollar amounts so determined are defined as "earnings."

From the contractor's point of view, however, the wage rate for an employee working under a typical union contract includes the base wage rate plus the hourly rates for all fringe benefits. As discussed in Chapter 3, all unit costs for labor and the estimated costs for all job items are based on wage rates so defined.

The hourly rates for each fringe benefit are multiplied by the hours worked, and these amounts are paid into trust funds set up by the union for each fringe benefit. The rates used for fringe benefits are the same regardless of whether the hours worked are at straight time or overtime. Typical fringe benefits are:

1. Vacation plans.
2. Pension plans.
3. Health and welfare plans.
4. Other minor plans such as apprenticeship training or industry promotion.

On nonunion, prevailing-wage contracts, all of the fringe benefits are paid directly to the workmen. Premium pay is normally paid for overtime work, and the premium pay rate is a percentage of the base wage rate. Thus, when time and one half is paid for overtime, the rate of premium pay will be 50 percent of the base wage rate, and when double time is paid, the rate of premium pay will be 100 percent of the base wage rate.

The typical components of the payroll expense item are discussed in the next section, and a method is developed whereby the specified percentage rates for each payroll expense, based on earnings, are converted to a single, composite rate applicable to wages. This composite rate is then applied to the total estimated labor cost for each job item.

2.2.1 Payroll Taxes

Payroll taxes normally consist of federal social security taxes (Federal Insurance Contributions Act, or FICA), federal unemployment taxes, and state unemployment taxes. The specified percentage to be charged for each of these taxes is applied to a dollar amount based on the base wage rate plus the vacation plan rate plus the premium pay rate, when applicable.

Assume the following conditions:

Base wage rate: $10.00 per hour
Vacation plan rate: $1.50 per hour
Overtime: Time and one half
Straight time hours: 40
Overtime hours: 6
Total hours worked: 46

Then the premium pay rate for overtime would be 50 percent × $10.00 = $5.00, and the dollar amount to which the percentages for taxes would be applied would be 46 × ($10.00 + $1.50) + 6 × $5.00 = $559,00.

In the following discussion of the individual payroll taxes, the dollar amount calculated as above is referred to as "earnings including premium pay."

2.2.1.1 FEDERAL SOCIAL SECURITY TAXES (FICA). Only that portion of the social security tax which is paid by the employer is considered in determining the allowance for payroll expenses.

Each employee also pays a portion of the social security tax. This portion is withheld by the employer and is paid to the Internal Revenue Service for credit to the employee's social security account. Except for the overhead involved, the tax paid by the employee does not constitute an additional expense to the contractor.

In our estimates, we assume that the employer's portion of the social security tax is 5.85 percent of the first $16,500 of earnings, including premium pay, paid to each employee.*

2.2.1.2 FEDERAL UNEMPLOYMENT TAX. In our estimates, we assume this tax to be 3.2 percent of the first $4200 of earnings,† including premium pay, paid to each employee and assume further that the 3.2 percent rate may be reduced by 90 percent of the rate used in determining the state unemployment tax, up to a maximum deduction of 2.7 percent.

2.2.1.3 STATE UNEMPLOYMENT TAX. In our estimates we assume this tax to be 3.6 percent of the first $7000 of earnings, including premium pay, paid to each employee.

2.2.2 Payroll Insurance

The costs for workmen's compensation insurance and for all or part of the contractor's liability insurance are determined by applying specified percentage rates to the employee's earnings.

*1979 rate of 6.13 percent of the first $22,900 is used for example only.
†1979 rate of 3.4 percent of the first $6000 is used for example only.

2.2.2.1 WORKMEN'S OR WORKERS' COMPENSATION INSURANCE. This insurance covers medical and other costs resulting from job-related accidents or afflictions. The cost for this insurance has increased drastically in the past few years, and it now represents the major portion of the payroll expenses.

Workmen's compensation insurance is mandated by the states; their procedures for assessing its cost may vary, and the percentage rates charged are subject to frequent changes. The applicable rates and procedures must, therefore, be obtained from the insurer in all cases. In the following discussion, we assume the job is located in California.

The percentage rates used to determine the costs for this insurance vary greatly, on a given job, depending on the particular classification of work being performed. The actual costs are determined by an audit of the contractor's books, and it is essential that the work classifications for the workmen be noted on the daily time cards.

The percentage rates for workmen's compensation insurance are applied to a dollar amount based on the base wage rate plus the vacation plan rate, and premium pay for overtime work is not included. Thus referring to the example given in Section 2.2.1, the dollar amount to which the percentages for workmen's compensation insurance would be applied is

$$46 \times (\$10.00 + \$1.50) = \$529.00$$

The dollar amount so calculated will be referred to as ''earnings excluding premium pay.''

In our estimates, we assume the following percentage rates for workmen's compensation insurance:

Classification	Percentage of Earnings Excluding Premium Pay
Concrete work NOC (not otherwise classified)	12.5
Concrete bridges and culverts	28.5
Demolition work	30.0
Carpentry work	15.0
Earthwork and roadwork	6.5

2.2.2.2 LIABILITY INSURANCE. The several types of liability insurance normally carried by contractors were discussed in Section 2.1.1. In our estimates we assume a composite rate of 2.8 percent for the following insurances:

1. Public liability and property damage (including automobiles).
2. Contingent liability and property damage insurance.
3. Products and completed operations liability insurance.
4. Excess limits insurance.

The 2.8 percent rate is usually applied to earnings, including premium pay, as defined previously in the section on payroll taxes. Insurance rates and terms will vary, however, with the insurer used. In all cases, the specifications must be checked for special requirements which might affect the rates or the limits for the insurance.

2.2.3 The Allowance for Payroll Expenses

The contractor's cost estimates for each job item are based on wage rates which include the base wage rate plus all fringe benefits. As discussed in the preceding, the specified percentage rates for payroll taxes and insurances are applicable to earnings which are based on the base wage rate plus only the vacation plan. Premium pay for overtime is included in some instances and excluded in computing the cost of workmen's compensation insurance; all earnings in excess of certain stated amounts are excluded in determining the costs for payroll taxes.

The final costs for all of the payroll expense items are determined by audits. For estimating purposes, however, the contractor must determine a single composite percentage rate for all payroll expense items, and this rate must be applicable to the total labor costs shown on the estimate sheets for each job item.

Based on the assumed percentage rates and limits previously outlined for payroll taxes and insurances, a reasonable approximation of the composite percentage rate for payroll expenses is easily determined. Although state taxes will be different for the various states and the rates for all taxes and insurances tend to increase with the passage of time, the method for determining the composite rate for payroll expenses, outlined in the following example, will be applicable in all cases.

The assumed hourly earnings and wage rates are shown in Table 2.1, and these rates are used in the cost estimates presented in Chapter 8.

The total yearly earnings shown in Table 2.2 are based on the assumption that the average hourly-paid workman will be in our employ for only 3 months, or 520 hours, during the year, whereas the superintendent, a salaried employee, will work all year long. This assumption is reasonable for a beginning contractor bidding his

TABLE 2.1   Earnings and Wages in Dollars per Hour

Classification	Base Wage Rate	Fringe Benefits				Earnings[a]	Total Wages[b]
		Vacation	Pension	Health and Welfare	Other		
Carpenter	$11.69	$1.00	$1.95	$1.60	$0.10	$12.69	$16.34
Laborer	9.28	0.70	2.20	1.20	0.10	9.98	13.48
Cement finisher	11.50	1.25	2.00	1.25	0.10	12.75	16.10
Superintendent	13.00	0.75	—	—	—	13.75	13.75

[a]Earnings = base wage rate + vacation plan rate.
[b]Total wages = base wage rate + all fringe benefits.

TABLE 2.2 Payroll Taxes Paid

Classification	Carpenter	Laborer	Cement Finisher	Superintendent
Total yearly earnings (in our employ)	$6600	$5200	$6600	$28,600
Type of tax			Taxes Paid	
FICA[a]	$ 386	$ 304	$ 386	$ 965
FUI[b]	$ 21	$ 21	$ 21	$ 21
SUI[c]	$ 238	$ 187	$ 238	$ 252
Total taxes	$ 645	$ 512	$ 645	$ 1238
Taxes as a percentage of yearly earnings	9.8%	9.9%	9.8%	4.3%

[a]5.85% of first $16,500.
[b](3.2 − 2.7)% of first $4200.
[c]3.6% of first $7000.

first small job. Established contractors doing larger jobs may find that their average workman is employed by them for 8 to 10 months during the year, and this will reduce the taxes expressed as a percentage of earnings, in Table 2.2, by as much as 3 percent.

Established contractors, however, keep cost data showing the total wages paid during each past year and the total costs for each of the payroll taxes. The percentages, based on wages, paid for payroll taxes in the past year are then determined directly and are adjusted to fit the rates and limits applicable to the current year.

In Table 2.3 the specified percentage rates for payroll taxes and insurances are converted to rates applicable to the estimated labor costs. The factors used for the conversion are taken from Table 2.1 and are the ratio of the hourly earnings to the hourly wages for each classification of employees. The percentage rate for workmen's compensation insurance will vary depending on the type of work being done.

TABLE 2.3 Payroll Expenses Expressed as Percentage of Estimated Labor Costs

Classification	Carpenter	Laborer	Cement Finisher	Superintendent
Taxes	9.8%	9.9%	9.8%	4.3%
Compensation insurance	12.5%	12.5%	12.5%	12.5%
Liability insurance	2.8%	2.8%	2.8%	2.8%
Total percentage based on earnings[a]	25.1%	25.2%	25.1%	19.6%
	12.69	9.98	12.75	1
Factor	16.34	13.48	16.10	
Total percentage based on wages[b]	19.5%	18.7%	19.9%	19.6%

[a]Earnings = base wage rate + vacation plan rate.
[b]Wages = base wage rate + all fringe benefits.

The rate for concrete work NOC is used for illustration in Table 2.3. Since the cost for workmen's compensation insurance now constitutes a major portion of the payroll expense item, and since the rates charged for this insurance vary widely for different classifications of work, the total percentage to be allowed for payroll expenses must be calculated for each classification of work that the contractor will perform.

Assuming the composition of a typical crew to be six carpenters plus three laborers plus one cement finisher, the percentage to be allowed for payroll expenses for concrete work NOC will be

$$(6 \times 19.5 + 3 \times 18.7 + 1 \times 19.9) \div 10 = 19.3 \text{ percent}$$

and this percentage is applied to the total labor cost shown on the estimate sheets for such work. In view of the assumptions required to determine this percentage rate, it would, in practice, be rounded up to 19.5 percent. It is essential that the actual costs for all of the payroll expense items be monitored, on an on-going basis, against the amounts recovered by the application of the percentage rate used.

In the preceding example, no allowance was made for the effect of premium pay for overtime even though the costs for normal overtime are reflected in the unit costs and will appear in estimated costs for the job items. The amount of overtime worked is normally small with respect to the total hours worked, and its effect can be neglected in all typical situations.

2.3 GENERAL OVERHEAD AND PROFIT

General overhead is the cost of running the business as a whole and is not directly related to the performance of any specific job. General overhead includes the cost of the following items:

1. Owner's salary.
2. Home office salaries.
3. Home office rent.
4. Home office power and water.
5. Home office telephone and radios.
6. Home office vehicles.
7. Home office supplies and repairs.
8. Depreciation.
9. Contractor's yard.
10. Professional services.
11. Business taxes and licenses.
12. Insurance.
13. Health insurance for salaried personnel.

14. Dues and subscriptions.
15. Travel and entertainment.
16. Bidding expenses.
17. Payroll expenses on general overhead labor.
18. Donations.
19. Special government programs.

The monthly or yearly cost of the contractor's general overhead is easily estimated, but the amount to be charged to any given job depends upon the total volume of work done during the course of the year.

The contractor has no way to predict precisely the volume of work that will be awarded to him during an upcoming year. This problem is common to all business enterprises but is, perhaps, more severe in the construction industry; the sizes of the jobs bid vary greatly, the number of jobs awarded per year is relatively small, and a difference of one job of average size will often result in a significant variation in the volume of work done during the year.

In practice, the contractor decides on an amount to be added to the bare cost of each job for general overhead and profit and then sets a goal to obtain sufficient work to pay the general overhead and to make a profit over a year's time.

The bare cost of any job is the total cost of doing the work including direct job overhead but does not include the allowance for general overhead and profit. The amount allowed for general overhead and profit is the contractor's markup; it may be expressed in dollars or as a percentage of the bare cost of the job.

The amount of the contractor's markup is limited by the competition from fellow contractors and will decrease as the number of bidders increases.

Preliminary cost estimates prepared by architects and engineers frequently contain an item for contingencies. Except in the case of rare and unusual circumstances, the contractor never includes an allowance for contingencies as a separate item of cost in bidding work. To include such an item would result in a high bid in most cases. Each job item is bid as the estimator sees it, and the costs of all conditions that can reasonably be expected to occur are included in the estimated costs of the items involved.

Those increased costs that would result if unlikely events were to occur are considered by the contractor at the time the size of the markup is determined. Where such costs represent a small percentage of the normal markup, they are usually ignored. Where such costs equal or exceed the normal markup, the markup may be increased, and the amount of the increase depends on the contractor's judgment of the probabilities involved, his need for the particular job, and the relationship between the possible increased cost and his overall financial condition.

The bare cost of a completed project will never exactly equal its estimated bare cost, and there are no guarantees the contractor will make a profit on any job awarded to him. Because of the inherent uncertainties of the estimating process, there are practical lower limits to the amount of the markup used; these limits should not be violated often if the contractor plans to stay in business. A theoretical analysis of past bidding history may indicate more money could be made by reducing the markup and thereby doing more work. The contractor, however, runs considerable risk when the markup used is below the practical lower limit for the type of work involved.

In the final analysis, the size of the markup will depend on the contractor's feeling for the job, and this feeling will be influenced by a number of considerations. Included among these will be his need for the particular job at the time, the amount of other work available, the number of other bidders, his past bidding experience on similar work against particular competitors, his past cost history on similar work, the type, size, and location of the job, the risks involved, and the character of the owner and his representative based on past experience or inferred from the plans and specifications. In all cases the markup used should be considered both as a lump sum and on a dollar-per-month-of-job-duration basis.

In our estimates we allow 10 percent for general overhead and profit on engineering work.

On building work the general contractor does about 25 to 30 percent of the work with his own forces, the uncertainties are fewer, and the markup is smaller. As rules of thumb, the following three parameters may be considered in determining a reasonable markup for building jobs.

1. Twenty-five percent of the total bare cost of the general contractor's own work. This cost is taken directly from the estimate summary sheet and includes the contractor's costs for labor, payroll expenses, materials, and rentals. The costs for direct job overhead are included.

2. Forty percent of the total labor costs for the general contractor's own forces. This cost includes direct job overhead labor and all payroll expenses and is the sum of the labor costs shown on the estimate sheets for the contractor's own items of work.

3. Five to seven percent of the estimated bare cost of the job. The bare job cost equals the total job cost less the contractor's markup for general overhead and profit and is shown on the estimate markup for general overhead and profit and is shown on the estimate summary sheet. The upper limit is considered when the general contractor will do a larger percentage of the work with his own forces, as is usual when concrete buildings are involved. The lower limit is considered when more of the work is done by subcontractors, when the contractor has good cost records for the major categories of work required, and when no particular hazards are foreseen.

CHAPTER 3

Unit Costs

Contractors engaged in repetitive work find it to their advantage to set up a system of unit costs for each job item. When labor, material, or equipment rates change, the unit costs can be updated with ease, as indicated in the following paragraphs.

The total costs for the labor and equipment required for each job item are estimated by applying unit costs for labor and equipment to the quantities of those primary activities needed to complete the item. The unit costs are directly proportional to the hourly wage and equipment rental rates. The wage rates used include the costs of all fringe benefits, and the rental rates used include the costs of operation and maintenance together with the costs of subsistence payments to the operators when required.

The costs for labor must be separated from the costs for materials and rentals on the estimate sheets. This is mandatory because certain payroll expenses, discussed in Chapter 2, must be applied to the total labor cost of each job item as a percentage of that labor cost. It is therefore necessary to use a unit cost for labor and a separate unit cost for equipment, when required.

3.1 UNIT COSTS AND PRODUCTION RATES

Production may be expressed directly in terms of production rates or indirectly in terms of unit costs. Numerically, the two terms are inversely proportional to one another, but in the broadest sense, they are interchangeable, and either is easily derived from the other. The unit costs or the production rates for a given primary activity will exhibit wide variation depending on the type of work being done.

For example, the unit cost to fine grade an abutment foundation will be about 3.5 times the unit cost to fine grade a grade slab, and these costs reflect the relative magnitudes of the quantities normally encountered, the typical subgrade conditions prior to commencing the fine grading operation, and the sizes of the crews normally used.

Similarly, a dragline will perform mass excavation at 3 to 3.5 times the rate achieved when excavating for bridge abutments, and the lower rate reflects the smaller quantities involved and the constant care required to maintain lines and grades.

Therefore, unit costs or production rates must be kept for primary activities relating to the various types of work encountered, and such costs or rates will, in a general way, reflect the typical conditions relating to the particular operation. In all cases, production rates must be indexed to a particular crew size or to a particular piece of equipment, and unit costs must be indexed to the applicable wage or rental rates.

The relationship between unit costs and production rates may be expressed as follows:

$$R = \frac{C}{U} \tag{3.1}$$

or

$$U = \frac{C}{R} \tag{3.1a}$$

where

R = production rate, per hour or per day

U = unit cost

C = cost of crew or equipment per hour or per day

Equation 3.1 shows that to convert a unit cost into a production rate, a crew or equipment size must be assumed and that higher production rates will be achieved when larger crews or equipment are used. This relationship is only true within certain practical limits.

To obtain a job, the contractor must bid it at the lowest possible cost, and this requires the balancing of two conflicting requirements. On the one hand, the direct job overhead costs, discussed in Chapter 2, will decrease as the time required for completion is reduced. This aspect, considered separately, seems to indicate the use of large crews. On the other hand, minimum unit costs can only be achieved when the crew and equipment sizes are in proper relationship to the quantities of work to be done.

The job superintendent is responsible for doing the work at the lowest possible cost, and he determines the sizes of the crews and equipment actually used. The unit costs and production rates developed on past work will, in general, reflect optimum crew and equipment sizes. For lowest unit costs, small crews and equipment are used when small quantities are involved since costs decrease when daily repetitive operations can be established. The proper selection of crew and equipment sizes is critically important, and the ability to make this selection can only be acquired through on-the-job experience.

Production rates are generally used in connection with simple job activities and must be used when job conditions, which directly affect the resulting unit costs, vary greatly from job to job or from one part of a job to another.

Mass scraper excavation is a simple activity even though the scraper excavation may be a major item; no secondary activities are involved, the only variable being the round-trip cycle time for the scrapers, although the haul distances and the resulting cycle times are subject to wide variation.

Unit costs are generally used in connection with complex activities where the realized unit costs are not greatly affected by the varying conditions found on different jobs.

Since construction cost estimates are expressed in dollars and since time-of-completion estimates are based, in general, on the costs to do certain portions of the work, unit costs are always used on the estimate sheets. Production rates, when used, are converted to unit costs on a work sheet kept immediately behind the estimate sheet in question.

Unit costs based on past cost records will reflect the costs for all incidental overtime, the additional costs for foremen and for different classifications of workmen, and the relative magnitudes of the quantities involved. Overtime costs that can be foreseen, such as the costs to cure concrete over weekends, are estimated separately as a part of the item requiring the overtime.

Job payrolls, cost records, and bank statements are easily reconciled, and since the quantities have been taken off, direct unit costs are readily determined for work under construction.

By comparison, the determination of accurate production rates from job cost records is extremely cumbersome when complex activities are involved. Separate records of the accumulated hours for each classification of workmen must be kept for each primary activity, and these hours must be converted to dollars for purposes of reconciliation.

3.2 ACCURACY OF THE COST DATA

The contractor's survival depends, literally, on the accuracy of his cost data. One of management's prime responsibilities is to ensure that cost records for work in progress are kept accurately. On a well-organized job, all costs are charged to the accounts to which they belong, and the chips fall where they may. It is impossible to take effective corrective measures if the contractor is unaware of where the trouble lies.

Construction firms have experienced difficulties when dishonest superintendents have falsified cost data. Typically, such superintendents tend to charge costs properly belonging to an overrunning item for which they are solely responsible to other large job items where some of the responsibilities are shared by management or by a joint venture partner. When an untested superintendent has an item overrunning the estimate, management must investigate to determine if the trouble is really where the cost records show it to be.

3.3   OVERALL UNIT COSTS

Overall unit costs are obtained by dividing the total cost for an entire job item by the number of units in the item. Overall unit costs are used to prepare preliminary estimates and to furnish a rough check when detailed cost estimates are made.

Overall unit costs are inherently imprecise since the unit costs for the constituent parts of an item and the relative size of these parts will vary. In the preparation of detailed construction cost estimates, overall unit costs may be used in connection with minor or insignificant items.

3.4   UNIT COSTS FOR PRIMARY ACTIVITIES

For maximum precision, unit costs are applied to those primary activities which must be performed to complete each constituent part of the item. The unit cost for a particular primary activity will include the costs for those related secondary activities which must normally be performed to complete the operation. The primary activities are selected so that the relative amounts of the related secondary activities are reasonably uniform from job to job. Combining the costs of the secondary and primary activities greatly reduces the number of activities for which cost records must be kept, simplifies the takeoffs and the estimate sheets, and reduces the likelihood of error.

From the general contractor's viewpoint, the primary and secondary activities relating to the construction of reinforced concrete bridge abutments are:

1. Fine grade.
2. Set and strip foundation forms.
   a. Oil forms.
   b. Set and strip starter walls, solid keyways, and templates for reinforcing steel.
   c. Place incidental amounts of expansion joint material.
   d. Move forms ahead to the next pour.
3. Pour foundation slab.
   a. Prepare for pour.
   b. Clean up after the pour.
   c. Cure during regular working hours.
4. Make abutment and wingwall form panels.

5. Set and strip abutment and wingwall form panels.

   a. Oil forms.

   b. Set and strip chamfer and pour strips.

   c. Place incidental amounts of expansion joint material.

   d. Move forms ahead to the next pour.

6. Pour abutments and wingwalls.

   a. Prepare for pour.

   b. Clean up after the pour.

   c. Cure during regular working hours.

In addition to these activities, some abutments may require weepholes, utility openings, or the setting of anchor bolts. Such activities are not required on all jobs, their quantities will vary widely, and they are treated, therefore, as primary activities to which separate unit costs are applied.

Making wall form panels is considered a primary activity since the quantity to be made will depend on the number of reuses which are possible and may not be directly related to the total wall areas. Methods of determining form reuses are given in Chapter 8.

3.5 UNITS OF MEASURE

The units of measure used for unit costs must relate directly to the amount of work to be done. Most labor and equipment costs are directly proportional to the weight of the material to be handled. When this weight is proportional to area, unit costs are, for convenience, often based on areas. Thus, the unit costs for setting and stripping formwork are expressed in dollars per square foot, and the unit costs for wall formwork, presented in Chapter 8, assume $\frac{5}{8}$-inch plyform, 2 × 4 studs, double 2 × 4 wales, and taper ties at 2 feet on center both ways.

Similarly, the cost to set and strip bridge falsework depends on the total weight of material handled, and the unit cost for this work increases with the height of the falsework. Unit costs for setting and stripping wood falsework are expressed in dollars per thousand board feet for falsework of a given height, and the unit costs for setting and stripping steel shoring towers are expressed in dollars per hundred-weight of material handled for towers a given number of frames in height.

Such unit costs exhibit reasonable consistency from one job to another. If the unit costs for falsework were based on the square feet of soffit form supported, no valid comparisons would be possible, except in the case of nearly identical structures.

In all cases, the amount of bridge falsework required is based on a quick preliminary design. This procedure automatically accounts for the additional falsework required under skewed bridges and when roadways must pass through the falsework.

3.6 ACCURACY OF THE ESTIMATE

Although the completed estimate will be mathematically correct, the estimation of construction costs is not and can never be an exact science; the final selection of the unit costs used will, in every case, be based on the estimator's experience and judgment.

The unit costs realized on past work can be determined with a fair degree of precision, but it is impossible to predict the exact unit cost that will be realized on a proposed project. The cost to do work is subject to many variables; from week to week, on the same job, the unit costs for identical work, by the same crews, will vary by 5 percent or more. Among the factors affecting unit costs are:

1. The work methods used.
2. The amount of access available.
3. The quantity of the work, its complexity, and the opportunity to establish repetitive operations.
4. The size of the crews or equipment used in relation to the quantity of work to be done.
5. The availability of materials.
6. The weather and soil conditions.
7. The quality of workmanship required.
8. The competence of the job inspection and supervision.
9. The availability of competent workmen and of the needed equipment.
10. The degree of cost control exercised.
11. Interruptions by strikes.
12. Union rules.
13. Groundwater and water control.

Although the experienced estimator is well aware of the inherent imprecision in selecting unit costs, the selection is based on judgment and cost data, not on guesswork. In the broadest sense, all unit costs and production rates are selected by comparison.

The experienced contractor will have extensive cost records showing his unit costs and/or production rates for the primary activities relating to all of his past work, and his livelihood depends on the completeness and the accuracy of these data.

These unit costs will relate to the particular job conditions under which they were developed and to the contractor's cost system. The unit costs for a particular primary activity from several different jobs may show a maximum variation of as much as 100 percent. The reasons for the extreme variation are usually evident and reflect unusually small quantities of work in some cases, severe job conditions in others, and sometimes, poor cost data.

When an array of past unit costs for a given primary activity is considered, a probable value for the average unit cost applicable to average conditions can be selected, and the variation for unit costs relating to average conditions will normally range from 5 to 10 percent. The true worth of the experienced estimator lies in his ability to select a reasonable unit cost from the array, consider the job conditions to which it applies, and modify it, as necessary, for use on a proposed project with its particular job conditions.

In all cases, extreme care must be exercised when large quantities are involved.

Some individuals in the construction industry decry the use of unit costs. These individuals have one thing in common; they have never personally made a detailed construction cost estimate.

The use of unit costs is the only rational method for estimating construction costs. The procedure must be based on skill, experience, cost data, and judgment; it can never be precise, but use is made of all available data and information. The thousands of successful contractors are solid evidence that the procedure works.

3.7 WARNING REGARDING UNIT COSTS

Each contractor's unit costs and production rates are unique, and all unit costs and production rates are dangerous if indiscriminately used.

In the following chapters, average unit costs and production rates are presented for many types of work performed by general contractors. These costs assume average job conditions for work done in a temperate climate under the continuous supervision of a good superintendent. These unit costs are shown to illustrate the procedures followed in estimating construction costs.

It is virtually impossible to list every secondary activity whose cost is included or to enumerate all the typical job conditions under which the unit costs were realized. The use of the unit costs presented is illustrated, however, in many examples, all of which pertain to typical job situations.

The unit costs presented herein should not be blindly used by inexperienced estimators. When large quantities are involved, the unit costs used should, in every case, be based on the estimator's personal cost records for similar work performed under known job conditions. When moderate quantities are involved, the unit costs presented herein should only be used by estimators who have the experience and background to confirm their magnitude by consideration of the estimator's own cost data for similar work.

3.8 INDEXING UNIT COSTS

Unit costs developed on past jobs having various wage and equipment rental rates must be converted to be applicable to a proposed project having different rates.

The conversion process is simple.

A series of indexes is assigned to all jobs past or proposed, and as illustrated in the estimates in Chapter 8, the indexes used are written at the top of the estimate sheets for each job item. The labor index is the cost of an assumed crew per hour or per day, and the equipment rental indexes are the hourly rental rates for the equipment. The unit costs are directly proportional to the applicable indexes.

In the following chapters, average unit costs and their related indexes are given for many types of work performed by general contractors. In Chapter 8, for example, the indexes are based on the following wage and rental rates, and the wage rates include all fringe benefits:

Carpenter	$16.34 per hour
Laborer	$13.48 per hour
Cement finisher	$16.10 per hour
35-ton truck crane	$75.00 per hour

The labor index used for setting and stripping formwork is the cost of a crew of six carpenters plus two laborers working for 8 hours and is designated K_{6+2}. Based on the preceding rates, $K_{6+2} = 1000$. For a proposed job where $K_{6+2} = 1100$, those unit costs based on $K_{6+2} = 1000$ would be multiplied by $\frac{1100}{1000} = 1.10$. Similarly, unit costs from a past job indexed to $K_{6+2} = 700$ would be multiplied by $\frac{1100}{700} = 1.57$.

Where, with the passing years, the wage rates for carpenters and laborers increase at approximately equal percentage rates, the K_{6+2} index will convert unit costs with no error in any case. Where one of the two wage rates increases at an appreciably faster rate than the other, a small error will be introduced into the converted unit costs over a period of years in those cases where the actual composition of the crews used is appreciably different from the six carpenters plus two laborers assumed. This error should not exceed 2 percent over a period of 8 or 10 years and in all likelihood would be much less. In practice, unit prices are being determined and updated continuously, and the contractor will give more weight to his more recent cost records. In such cases the error from this source will be negligible.

Where work is primarily done with laborers the K_L index, which is the hourly wage rate for one laborer, is used. The unit costs presented herein for such work are indexed to $K_L = 13.48$.

Where work is done with laborers and cement finishers in approximately equal numbers, the K_{L+F} index is used, which is the hourly wage rate of one laborer plus one cement finisher. The unit costs for such work presented herein are indexed to $K_{L+F} = 29.58$.

The wage rate for a given trade may vary with the type or classification of the work being performed. Thus a vibrator operator may receive more pay while vibrating than when he is performing as a general construction laborer. In such cases, the wage rate used in indexing unit costs is the prevailing rate for the majority of the workmen involved, and the premium paid to those engaged in special activities will be automatically included in the unit costs developed.

The unit costs for concrete-placing equipment presented in Chapter 8 are indexed to $K_{Cr} = 75$, which is the hourly rental rate for a 35-ton truck crane, operated and maintained. The equipment costs for a number of miscellaneous activities are indexed to K_{BH}, which is the hourly rental rate for a small rubber-tired loader-backhoe. The costs to place concrete with concrete pumps are discussed in Section 8.2.6.1.

3.8.1 On-and-Off Charges for Equipment

The unit cost for doing work with equipment depends upon the hourly rental rate charged, the production rate achieved, and the costs to move the equipment on and off the job. Ideally, on-and-off charges should be estimated separately, and this is done whenever the number of on-and-off moves can be predicted with reasonable accuracy. Thus the costs to move earth-moving equipment on-and-off the job are always considered as a separate item of expense.

On the other hand, when equipment is required to place concrete, the number of on-and-off moves required is often difficult to predict, and the unit costs for this work, presented in Chapter 8, include 1 hour of travel time. As noted on the unit cost tables, these unit costs assume pours of average size for the type of work being considered.

3.8.2 Increase in Unit Costs with the Passage of Time

Increases in unit costs due to rising wage and rental rates are easily determined, as already outlined. However, unit costs for a given activity will also slowly increase with the passage of time because of the decreasing productivity of the workmen. Such cost increases tend to be slow and irregular, cannot be predicted with much accuracy, and in most cases are within the limits of accuracy of the unit cost selection process. In practice, the estimator places more weight on his more recent unit costs, and this procedure accounts for the decreasing productivity of the crews.

3.9 CONVERTING UNIT COSTS TO METRIC UNITS

English and metric units are readily converted one to the other. When making such conversions, it is essential to differentiate between sentences and equations. Thus, "1 inch = 2.54 centimeters" is a correct statement or sentence. The corresponding equation, however, is

$$\text{inches} \times 2.54 = \text{centimeters}$$

To avoid confusion, the following conversion factors are expressed as equations:

Inches	$\times 2.54001$	= centimeters
Feet	$\times 0.304801$	= meters
Yards	$\times 0.914402$	= meters

Square inches	× 6.45165 =	square centimeters
Square feet	× 0.0929034 =	square meters
Square yards	× 0.836127 =	square meters
Cubic inches	× 16.3873 =	cubic centimeters
Cubic feet	× 0.0283172 =	cubic meters
Cubic yards	× 0.764559 =	cubic meters
MFBM	× 2.35975 =	cubic meters
Pounds	× 0.453592 =	kilograms
Tons (2000 pounds)	× 907.185 =	kilograms
Centimeters	× 0.393699 =	inches
Meters	× 3.28083 =	feet
Meters	× 1.09361 =	yards
Square centimeters	× 0.155000 =	square inches
Square meters	× 10.7639 =	square feet
Square meters	× 1.19599 =	square yards
Cubic centimeters	× 0.0610229 =	cubic inches
Cubic meters	× 35.3142 =	cubic feet
Cubic meters	× 1.30794 =	cubic yards
Cubic meters	× 0.423774 =	MFBM (thousand feet, board measure)
Kilograms	× 2.20462 =	pounds
Kilograms	× 0.00110231 =	tons (2000 pounds)

Unit costs expressed in dollars and English units are converted to unit costs expressed in dollars and metric units as follows:

Multiply Unit Costs in Dollars per	By	To Obtain Unit Costs in Dollars per
Foot	3.281	meter
Square foot	10.76	square meter
Square yard	1.196	square meter
Cubic foot	35.31	cubic meter
Cubic yard	1.308	cubic meter
Pound	2.205	kilogram
Ton (2000 pounds)	0.001102	kilogram
MFBM	0.4238	cubic meter

3.10 CONVERTING UNIT COSTS TO OTHER CURRENCIES

Unit costs expressed in dollars are easily converted to unit costs expressed in another currency if the wage rates, expressed in the other currency, are known:

$$C' = C_d \times \frac{K'}{K} \qquad (3.2)$$

where

C' = unit cost expressed in a foreign currency

C_d = unit cost expressed in dollars

K' = the appropriate unit cost index based on the foreign currency and wage rates

K = the appropriate unit cost index based on dollars

For example, assume that a unit cost of \$0.20 per square foot indexed to K_{6+2} = 1000 is to be converted to a unit cost expressed in francs per square meter and that the converted unit cost will be used in connection with work where the wages are paid in francs. Assume the wages in francs are:

Carpenters: 27.25 francs per hour

Laborers: 22.50 francs per hour

Then K'_{6+2} = 27.25 × 6 × 8 + 22.50 × 2 × 8 = 1668, and from Equation 3.2, we have

$$C' = (0.20 \times 10.76) \times \tfrac{1668}{1000} = 3.59 \text{ francs per square meter}$$

In this equation, the factor 10.76 is taken from the preceding tabulation, and it converts the unit cost expressed in dollars per square foot to one expressed in dollars per square meter.

When, for estimating purposes, unit costs are converted to other currencies, the conversion depends solely on the factors set forth in Equation 3.2. The actual monetary exchange rates for the currencies involved are not relevant, since the wage and rental rates paid in different countries will be different, even when these rates are expressed in terms of a common currency.

It is essential to realize that unit costs determined from cost records for work done in the United States are based on using American crews and methods while working with American specifications under particular climatic conditions. The estimator must therefore apply judgment to converted unit costs in order to correct for differences in the anticipated efficiencies of the crews, methods specified, and climate.

CHAPTER 4

![black bar]

Earthwork

Whenever earth is excavated, three basic operations are normally involved: the material must be excavated, it must be hauled, and it must be placed somewhere. Grades must always be made, and when embankments are to be constructed, the material must be watered and compacted.

On building work, the estimator decides what must be done, assumes a reasonable value for the side slopes, takes off the quantities, estimates their costs, and enters a lump sum cost for the earthwork on the estimate summary sheet.

The estimating of earthwork costs on unit price contracts requires more care since only those quantities within the specified pay limits will be paid for. The takeoff, therefore, must include both a check of the quantities shown on the engineer's estimate, which will be based on the pay limits, and a determination of the probable quantities to be handled based on reasonable side slopes. In addition, on unit price contracts, the costs of certain additional work may be specified to be included in the price bid for a particular item.

Job scheduling sometimes requires material to be temporarily stockpiled. In such cases, two excavation operations are required even though only one will be included in the pay quantities.

The water table must be below the bottom of the proposed excavation, and the costs to control groundwater or to lower the water table, when necessary, are carried as a separate section of the cost estimate for the earthwork item.

When well point systems are necessary, the design of the system is furnished by the suppliers of the required equipment. The suppliers will quote rental rates for the equipment and will furnish cost estimates covering the installation, operation, maintenance, and removal of the system. The installation of the system is supervised by the supplier, but the labor and equipment required to install and to remove it are furnished by the general contractor, who will also operate and maintain the system.

The cost to excavate material depends on its physical properties, the quantities to be excavated, the equipment used, the skill of the equipment operators, the depth of cut, the size of the work area in relation to the equipment used, and the quality of the supervision.

The optimum depth of cut for a digging bucket is that which will permit the bucket to be fully loaded in one pass and will vary with the size of the bucket and the hardness of the material being excavated. Scraper hauling units load best when making a 4- to 6-inch cut. Thus, a scraper with a 10-foot cutting edge and a capacity of 25 bank yards will require a minimum loading distance of about

$$\frac{25 \times 27}{0.42 \times 10} = 161 \text{ feet}$$

4.1 PHYSICAL PROPERTIES OF THE MATERIAL

The type of material to be excavated is a basic cost factor since soft materials load faster and swell less. Earthwork contractors become very proficient at moving those soils found in their particular areas, but extreme care must be exercised when estimating earthwork costs in areas where one is unfamiliar with the soils and when large quantities are involved.

4.1.1 Density

The unit weight of the typical, in-place bank material must be considered since, in some cases, the amount of material that can be loaded or hauled per cycle will be governed by the weight instead of the volume. In addition, the densities of the bank and of the compacted embankment materials are used to determine the grading factor, which is discussed in the following section.

The density of the bank material is determined by laboratory tests which give the dry density of the material and the percentage, by weight, of retained moisture. The in-place unit weight of the bank material will vary with the gradation of the material, the amount of moisture retained in it, and its natural relative compaction. The approximate in-place unit weights for commonly encountered materials are:

Material	Weight in Pounds per Bank Cubic Yard
Silty sand	2700
Sand	3100
Earth	3100
Gravel	3400
Clay	3400

The percentage of retained moisture in the bank material and the percentage of moisture required for maximum density are both considered when estimating the amount of added water required for compaction.

4.1.2 Shrinkage and the Grading Factor

Material that has been compacted to 90 percent of its maximum density is said to have a relative compaction of 90 percent. Current specifications normally require a relative compaction of 90 percent for embankments, and the resulting dry density is usually greater than the dry density of the bank material. In such cases, more than 1 cubic yard (CY) of cut will be required to make 1 CY of fill, and the material is said to shrink. This relationship, for a given material, is expressed by the grading factor, and in all cases

$$C = \frac{E}{G} \qquad (4.1)$$

where

G = the grading factor
C = the total volume of cut required to make embankment E
E = the volume of embankment

The grading factor is determined as follows:

$$G = \frac{D_B}{D_E} \qquad (4.2)$$

where

D_B = the dry, in-place density of the bank material
D_E = the average dry density of the embankment material on completion

Assuming a specified relative compaction for the embankment of 90 percent, D_E in Equation 4.2 will range from 90 to 100 percent of the maximum density for the material.

The average density of the completed embankment cannot be precisely determined in advance. Field tests for embankment densities are made in the top 1.5 feet of the embankment as the work progresses, and additional compaction is attained as loaded rubber-tired equipment rolls over the tested layer while placing the layers above.

Grading factors are given on the job plans, and the values used will depend on the judgment and the experience of the engineer. Typical values for the grading factor will range from 0.75 to 0.90 when embankments are made of earth to a relative compaction of 90 percent.

The grading factor is widely used by highway departments and contractors to adjust earthwork volumes. Many soils engineers and civil engineers in private practice, however, use a shrinkage factor that is the reciprocal of G. In such cases, Equation 4.1 becomes

$$C = ES$$

where

S = the shrinkage factor = $1/G$

When embankments are made of a mixture of blasted rock or boulders and earth, the average density of the embankment cannot be determined with much accuracy, and the grading factor will usually be greater than 1 and, at best, be an educated guess.

When high fills are constructed, the subgrade may subside under the added weight, and additional cut will be required to complete the fill. An allowance for subsidence is normally included in the embankment quantities given on the road profiles, and the amount of subsidence assumed will usually range from 0.15 to 0.30 feet. It is important to note that while the allowance for subsidence increases the cross-sectional areas of the embankment sections, subsidence does not affect the grading factor, which relates to shrinkage.

Large areas requiring shallow cuts and fills are often graded with wheel tractor-scrapers. Such equipment will compact the original ground surface during the grading operation, and the resulting subsidence may be as much as 0.4 foot. Since shallow cuts and fills over a large area are involved, this subsidence will cause a significant decrease in the amount of excavation otherwise available and will increase the amount of fill required in those cases where the grading plane was established without considering the subsidence.

To obtain an earthwork balance in such cases, the grading plane is often established so that the volume of excavation is as much as 50 percent greater than the volume of fill, and this percentage covers the combined effects of shrinkage and subsidence. The proper selection of the percentage to be allowed requires past job experience involving similar work and similar soils.

4.1.3 Swell and the Load Factor

When bank material is excavated and loaded, voids are created and the material is said to swell. This relationship is expressed by the load factor as follows:

$$L \times F = B \tag{4.3}$$

where

F = the load factor
B = the number of bank yards excavated
L = the number of loose (excavated) yards resulting from the excavation of B
    bank yards

and

$$F = \frac{D_L}{D_B} \tag{4.4}$$

where

D_L = the density of the loose material
D_B = the density of the bank material

Equation 4.4 may be restated as

$$F = \frac{1}{1 + S/100} \qquad (4.5)$$

where

S = the percentage of swell

and from Equation 4.5, we have

$$S = \left(\frac{1}{F} - 1\right) \times 100 \qquad (4.6)$$

Approximate load factors for commonly encountered materials are:

Material	Load Factor
Sand	0.90
Gravel	0.90
Earth	0.80
Clay	0.78
Hardpan	0.65
Blasted rock	0.60
Sandstone	0.60
Shale	0.60
Broken concrete	0.58
Stockpiled material	1.00

The load factor may be used to approximate the number of bank yards contained in a loader-scraper, truck, or bucket. Thus, if the load factor were 0.80, the number of bank yards contained in a 5-CY bucket would be 0.80 × 5 = 4.0 CY. In practice, estimating the number of bank yards carried per load is not that simple.

The load factor, or swell, taken from a table is an approximation, at best, and will vary in the same material depending on the method of excavating and loading. It may also vary by as much as 6 or 8 percent for similar pieces of equipment built by different manufacturers. In addition, earth-moving equipment may be rated at struck and at heaped capacity, and sound judgment is required to select that capacity to which a load factor should be applied.

On unit price contracts, an error in the load factor is more serious than an error in the grading factor. If the assumed grading factor were 5 percent too high, 5 percent more excavation would be required, and the haul lengths would change slightly. The job would end up short of dirt, and the additional excavation required would be paid for under a contract change order. The contractor would, however, receive his bid price for each yard excavated under the item as well as his negotiated price for each yard handled under the change order.

If, on the other hand, the assumed load factor were 5 percent too high, 5 percent more equipment cycles would be required to handle the contract amount of excavation, and no payment would be made for the additional work since the number of cubic yards excavated would not have changed.

In practice, the number of bank yards carried per load is based on experience and past job records for work done with similar equipment in similar types of soil, and the job records will include actual load counts relating to surveyed amounts of yardage.

4.1.4 Moisture Required for Compaction

The percentage of moisture by weight required for maximum density is determined from laboratory tests. Points are plotted to produce a curve having dry densities for ordinates and percentages of moisture for abscissas. As increasing amounts of moisture are added to successive samples, the compacted densities increase until maximum density is reached, after which the densities decrease as more moisture is added.

The moisture density curve is plotted so that the distance representing a difference of 1 pound of density on the vertical scale is the same as the distance representing a difference of 1 percent of moisture on the horizontal scale. The shape of the resulting curve has an important bearing on the relative ease of compacting the material on the job. The slope of the curve on either side of the maximum point will be 45 degrees or flatter for sandy soils, and their specified compaction can be achieved over a relatively wide range of moisture content.

The same slopes will usually be steeper than 45 degrees for silty soils and quite steep for soils containing 50 percent or more of material passing the number 200 sieve. The specified compaction for such soils can only be attained when the moisture content is very close to the optimum, and it is not uncommon for the compacted densities of silty soils to decrease from 7 to 10 pounds per cubic foot for a 1 percent increase above the optimum moisture content.

Since the amount of moisture cannot be precisely controlled on the job, it is clear that soils having steep moisture density curves will present more problems during the compaction operation. Discs are usually required to facilitate the compaction of such materials. Small amounts of additional water can easily be mixed with material layers that are too dry, and when a layer is too wet, the discs will aerate the material and shorten the time required for the drying-out process.

4.1.5 Gradation

Well-graded soils compact more easily than poorly graded ones. On rare occasions, a poorly graded, silty sand may be encountered, which cannot be compacted, under job conditions, to a relative compaction of 90 percent. Such soils may have a relatively high percentage of material retained between the No. 100 and the No. 140 sieves, and under a microscope this fraction will be roughly spherical in shape. In such cases, it may be necessary to blend a small amount of sand with the bank material to achieve the required compaction.

4.2 EARTH-MOVING EQUIPMENT

Production rates derived for earth-moving equipment are often based on the assumption that the equipment will work steadily, with no delays. Such rates are said to be based on a 60-minute hour. On all jobs, minor delays are inevitable, and for estimating purposes, effective rates based on 45- to 55-minute working hours are used. A production rate of 500 CY per 60-minute hour will result in an effective rate of

$$500 \times \frac{50}{60} = 417 \text{ CY per hour based on a 50-minute hour}$$

The minutes per working hour used by the contractor will depend on his past cost records for particular types of equipment and on the anticipated job conditions. Effective production rates for track-type equipment are often based on 50- to 55-minute hours, and 45- to 50-minute hours are often used for rubber-tired equipment. Some successful trucking contractors use a 55-minute hour for on-highway hauling.

4.2.1 Bulldozers

Bulldozers (Figure 4.1) are most effective when the haul lengths are 300 feet or less. As indicated by the production rates that follow (LCY = loose cubic yards) a difference of 50 feet in the average haul length has a significant effect on the production achieved.

APPROXIMATE BULLDOZER PRODUCTION:
TRACK TYPE, POWER SHIFT MACHINE, 300
FLYWHEEL HORSEPOWER, STRAIGHT BLADE

Haul distance (feet)	Base Production Rate (LCY per 60-minute hour)
50	800
100	525
150	375
200	300
300	200
400	175
500	150

Figure 4.1. A Caterpillar D9 bulldozer, parallelogram-type ripper (Meadow Construction Company, Bakersfield, California).

Production rates are, in general, proportional to the horsepower of the machine used. The preceding rates assume an average operator working on a level grade in average earth. For other conditions, the following approximate correction factors should be applied.

Condition	Factor
Loose stockpiled material	1.2
Ripped rock	0.7
Dry sand or silt, mud, or wet clay	0.8
Dozing up a 20% grade	0.6
Dozing down at 20% grade	1.2
Direct drive transmission	0.8
U Blade	1.2
Angle blade	0.65

Production rates will vary for different makes and models of equipment. When possible, rates from past jobs having similar conditions should be used. When such data are not available, production rates for a particular machine should be obtained from the manufacturer.[*]

We use the approximate production rates shown in the preceding and assume the following conditions:

Bulldozer	250 HP, direct drive
Material	Ripped rock
Grade	Level

---

[*]The Caterpillar Tractor Co. *Fundamentals of Earthmoving* and *Caterpillar Performance Handbook* deserve a place in every contractor's library.

Haul distance   200 feet, one way
Working hour   50 minutes

Then the estimated production rate is

$$300 \times \frac{250}{300} \times 0.8 \times 0.7 \times \frac{50}{60} = 117 \text{ LCY per hour}$$

Assuming a load factor for ripped rock of 0.60, this is equivalent to

$$117 \times 0.60 = 70 \text{ BCY per hour}$$

4.2.2   Rippers*

The production rate for a ripper will depend on the tractor horsepower and tractive capability and on the degree of consolidation, or rippability, of the material.

Clays, shales, cemented gravel, and hardpan are easily ripped. Rock formations may be sedimentary, igneous, or metamorphic, and of these, sedimentary rocks are most easily ripped. The rippability of sedimentary rock varies with its hardness, the degree of decomposition, the thickness of the strata, and the amount of fracturing. Unfractured strata more than about 12 inches thick are difficult to rip. In general, igneous and metamorphic rocks can only be ripped when they are fractured or decomposed.

4.2.2.1   REFRACTION SEISMIC ANALYSIS. On highway work, cuts must sometimes be made in areas where the topsoil covers one or more layers of rock. In such cases, test boring data on the job plans will show the depth to the rock in the cut areas. When the subgrade for the proposed excavation is below the rock layers, the rippability of the rock may be determined with reasonable accuracy by refraction seismic analysis.

Such analysis is based on the speed with which seismic waves travel through the different kinds of subsurface materials; these speeds range from 1000 feet per second through loose soil to about 20,000 feet per second through hard, solid rock.

The seismic equipment consists of a geophone connected to an electric cable to which steel striking plates are attached. The striking plates rest on the ground and are spaced along the cable at 10-foot intervals. The seismic waves are produced by striking the plates, in succession, with a sledge hammer. The time intervals between the hammer blows and the arrival of the waves at the geophone are recorded, and an analysis of these data indicates the depths, thicknesses, and rippabilities of the various subsurface layers.

---

*Most of the material in this section has been taken from the following two Caterpillar Tractor Company publications:

1. *Handbook of Ripping*, Sixth Edition
2. *Caterpillar Performance Handbook*, Edition 8.

Seismic surveys are typically made by geological exploration firms prior to the bid date, their reports are sold to the prospective bidders, and the price paid depends on the number of contractors using the information. Assuming a 410 flywheel horsepower tractor equipped with a single-shank, parallelogram-type ripper, rock is generally considered rippable when the seismic wave velocities are within the following limits (in feet per seconds):

Igneous rock	3000–7500
Metamorphic rock	3000–7500
Sedimentary rock	3000–8500

Ripper tooth penetration is essential if rock is to be ripped. Although seismic velocities for sedimentary rock may indicate probable rippability, the material may not be ripped effectively if the fractures and bedding joints do not allow tooth penetration.

4.2.2.2 RIPPER PRODUCTION. Accurate estimates of ripper production in relation to seismic wave velocities can only be made by contractors who have production records and seismic data covering past work with their equipment in their area. The Caterpillar Tractor Company has developed the following data, however, with the proviso that they be used with discretion and in conjunction with job experience in ripping work.

Velocity of Seismic Waves (feet per second)	Production (bank yards per 60-minute hour)
3000	1500–3000
4000	1000–2250
5000	500–1700
6000	250–1200

These data apply to a 410 flywheel horsepower tractor equipped with a single-shank, parallelogram-type ripper, and in general, ripping production is proportional to the tractor horsepower and tractive ability. The upper production rates shown are only applicable when ripping conditions are ideal; under adverse conditions, the lower limits should be used.

When the production rate falls below 150 to 200 bank yards per hour, production may sometimes be increased by as much as three or four times if a second tractor is used as a pusher.

Generally, the most efficient ripping is done at speeds of 1 to 1.5 miles per hour. When working in soft material, it is preferable to use 2 or 3 teeth rather than increase the speed.

When ripping soft material and when the passes are longer than about 125 feet, maximum production is achieved when successive passes are made in opposite directions. Assuming a ripping speed of 1 mile per hour (88 feet per minute) and allowing for a 0.5-minute delay per pass at the turns, an estimate of the probable production in such cases may be made as follows:

$$S = \frac{88L}{L + 44} \qquad (4.7)$$

where

S = effective travel speed in feet per minute

L = length of pass in feet

and

$$P = \frac{60 \times S \times W \times D \times K}{27} \qquad (4.8)$$

where

P = production rate in bank yards per 60-minute hour

W = width loosened per pass in feet

D = depth of penetration in feet

K = a constant based on job experience (the Caterpillar Tractor Co. recommends 0.80 to 0.90)

Assume the following conditions:

270 horsepower ripper with 1 tooth

  6 feet between passes

 2.5 feet of penetration

250 foot pass lengths

We have

$$S = \frac{88 \times 250}{294} = 75 \text{ feet per minute}$$

and

$$P = \frac{60 \times 75 \times 6 \times 2.5 \times 0.80}{27} = 2000 \text{ bank yards per 60-minute hour}$$

When the passes are shorter than about 125 feet or when required by job conditions, passes are made in one direction and the tractor backs up after each pass. Assuming a ripping speed of 88 feet per minute and a reverse speed of 250 feet per minute, the effective speed for the ripping pass is

$$S = \frac{L}{L/88 + L/250} = 65 \text{ feet per minute}$$

and this value should be used directly in Equation 4.8 to estimate production.

Heavy ripping in rock results in high repair costs. Ripper tips are expensive and might have a life of about 1 month to as little as 30 minutes under the worst conditions. Tip life, under severe conditions, is sometimes estimated as 10 hours. However, as many as 1000 hours have been obtained in shale. Due to accelerated wear, the hourly costs for rippers working in heavy rock will be about two times the normal cost for the equipment.

4.2.3 Track Loaders and Wheel Loaders

Loaders buckets are rated at heaped capacity, which assumes 2 : 1 slopes for the material above the struck capacity line when this line is parallel to the ground. Under job conditions, the struck line will not be exactly parallel to the ground, the bucket will swing and bob, and a carry or spill factor is applied to the rated capacity when estimating loader production.

4.2.3.1 BANK YARDS PER CYCLE. The bank yards carried in the bucket, per cycle, are

$$B = R \times L \times C$$

where

B = bank yards in the bucket
R = rated bucket capacity
L = load factor
C = carry factor

The Caterpillar Tractor Co.[*] recommends the following values for the carry factor.

TRACK- TYPE LOADER CARRY FACTORS
(FIGURE 4.2)

Material	Factor
Uniform aggregates	0.85–0.90
Mixed moist aggregates	0.95–1.00
Moist loam	1.00–1.10
Soil with boulders and roots	0.80–1.00
Cemented materials	0.85–0.95

*Caterpillar Tractor Co., *Fundamentals of Earthmoving,* July 1975.

Figure 4.2. Track-type loader (The Caterpillar Tractor Co.).

Figure 4.3. Loader-backhoe.

WHEEL-TYPE LOADER CARRY FACTORS
(FIGURES 4.3 AND 4.4)

Material	Factor
Uniform aggregates	
Sand	0.95–1.00
$\frac{1}{4}$ inch	0.85–0.90
$\frac{5}{8}$ inch	0.90–0.95
1 inch and over	0.85–0.90
Mixed moist aggregates	0.95–1.00
Blasted material	
Well blasted	0.80–0.85
Average	0.75–0.80
Slabs or blocks	0.60–0.65

When loaders are used in hard or rocky material, when the depth of cut is far from optimum, or when top production rates are required, the excavation is made with bulldozers, and the loaders work from stockpiles. As noted previously, the load factor for stockpiled materials is assumed to be 1.00 for estimating purposes.

Aggregates are usually paid for by the ton, whereas buckets are rated in cubic yards. The approximate densities for stockpiled aggregates are:

Stockpiled Material	Density (tons per CY)
Dry sand	1.35
Moist sand	1.45
Dry gravel	1.32
Dry sand and gravel	1.50
Wet sand and gravel	1.69

Figure 4.4. Caterpillar wheel loader (5M Company, Bakersfield, California).

When loaders are used to load trucks, the truck and the loader capacities should be balanced, when possible, so that the trucks may be fully loaded with a whole number of passes.

4.2.3.2 CYCLE TIMES. The base cycle time for loaders is the time required to fill the bucket, swing 60 to 90 degrees, dump the material, and return to the bank or pile. For estimating purposes, the base cycle times based on a 60-minute hour will be about as follows:

Track-type loaders   0.6 minute
Wheel-type loaders   0.5 minute

When a swing of 180 degrees is required, these base cycle times increase about 33 percent.

When the loader must carry the material to a dumping site, the travel time must be added to the preceding base cycle time. Average travel speeds for loaders based on a 60-minute hour are about as follows:

Loader	One-Way Distance (feet)	Speed (feet per minute)
Track type	50	200
	100	330
	200 plus	500
Wheel type	up to 500	370
	500 plus	800

To achieve 800 feet per minute, the wheel loader must have a smooth haul road.

4.2.4 Motor Graders

The production for motor graders (Figure 4.5) or blades is, in most cases, related to the areas or distances covered rather than to cubic yards of material. Thus, unit costs for motor grader work are normally based on square yards, or stations, per hour:

$$P_A = \frac{S \times 5280 \times W}{9 \times N} = \frac{587 \times S \times W}{N} \qquad (4.9)$$

where

P_A = production rate in square yards per 60-minute hour
S = travel speed in miles per hour
W = effective blade width in feet
N = number of passes required

or

$$P_S = \frac{S \times 52.80}{N} \qquad (4.10)$$

where

P_S = production rate in stations per 60-minute hour

For estimating purposes, an average speed of 4 to 4.5 miles per hour is often used, and the effective blade width is taken as 10 feet for finish work and 12 feet for rough work, such as mixing water with fill material during the compaction of embankments.

4.2.4.1 MISCELLANEOUS MOTOR GRADER COSTS. Roadside ditches approximately 2 feet deep with 4 : 1 side slopes were excavated complete with a motor grader. Indexed to K_L = 13.48 and to an hourly rental rate of $39 for the motor grader, the unit costs were:

Labor $0.45 per CY
Rental $1.30 per CY

These unit costs do not include disposing of the excavated material.

A motor grader was used to excavate, backfill, and make the final cleanup for 200 CY of curb and gutter. The curb and gutter section contained 0.0617 CY of concrete per linear foot, and the total length was 32.4 stations. Indexed to K_L = 13.48 and to an hourly rental rate of $39 for the motor grader, the unit costs for this work were:

Labor $ 0.55 per CY of concrete
or $ 3.40 per station

Figure 4.5. Caterpillar motor grader (J. L. Denio Inc., Bakersfield, California).

Rental $ 2.70 per CY of concrete
  or $16.70 per station

4.2.5 Tractor-Scrapers

Scrapers may be drawn by track-type tractors or by rubber-tired tractors (wheel tractors) (Figure 4.6). Wheel tractor-scrapers are discussed, with examples, in Chapter 6. The same principles apply to tractor-scrapers, but the maximum travel speed will be about 8 miles per hour on well-maintained haul roads and as low as about 4 miles per hour when haul roads are poor.

The loading time for tractor-scrapers is often taken at about 1 minute for small scrapers less than 12 CY in capacity and at 1.5 minutes for larger scrapers. For maximum production, a push-tractor may be required.

4.2.6 Compaction Equipment

The production rate for compacting equipment may be estimated as follows:

$$P = \frac{S \times 5280 \times W \times T}{27 \times N \times G} = \frac{196S \times W \times T}{N \times G} \qquad (4.11)$$

where

P = production rate in bank yards compacted per 60-minute hour
S = travel speed in miles per hour

Figure 4.6. Caterpillar D8 tractor with 12-CY scraper.

W = width of roller in feet
T = thickness of the compacted lift in feet
N = number of passes required
G = grading factor

The average travel speed will be about as follows:

Vibratory rollers (Figure 4.7)	2 MPH
Self-propelled sheepsfoot	6 MPH
Self-propelled pneumatic	7 MPH
Sheepsfoot towed by a track-type tractor (Figure 4.8)	3–4 MPH

Experience with the soils in question is required to estimate the number of passes necessary to achieve the specified compaction, but in most cases, from four to six passes are sufficient. As previously discussed, the ease of compaction varies with the gradation of the soil, its moisture content, and the uniformity of the soil moisture mixture. In most cases, soils compact more easily, on the job, when the moisture content is about 1 percent greater than the optimum.

Sheepsfoot rollers work well in sandy or in clay soils. Since they compact from the bottom up, the compacted thickness of the individual layers is limited to the length of the feet, which will range from 8 to 9 inches.

A double 5 × 5 sheepsfoot roller consists of two rollers side by side, each 5 feet in diameter by 5 feet wide, and is normally towed by a track-type tractor (Figure 4.9). Such a roller has an effective width of about 11 feet and will compact sandy loam material to a relative compaction of 90 percent at an effective rate of about 700 bank yards per hour when working on embankments placed with wheel tractor-scrapers.

Figure 4.7. *Vibratory sheepsfoot roller (Southwest Welding and Manufacturing Co., Alhambra, California).*

Figure 4.8. Caterpillar 815 compactor (The Caterpillar Tractor Co.).

Vibratory rollers work best in granular materials and are effective to a depth of 2 feet or more. A single vibratory sheepsfoot roller 5 feet in diameter by 6 feet wide will compact sandy loam material at an effective rate of about 400 bank yards per hour, and no assistance will be required from the hauling equipment.

Figure 4.9. Double 5 × 5 sheepsfoot roller.

TABLE 4.1   Estimated Dragline Production (BCY per 60-minute hour)

Material Class	Dragline Bucket Size (CY)												
	$\frac{3}{4}$	1	$1\frac{1}{4}$	$1\frac{1}{2}$	$1\frac{3}{4}$	2	$2\frac{1}{2}$	3	$3\frac{1}{4}$	4	$4\frac{1}{2}$	5	6
Moist loam or sandy clay	130	160	195	220	245	265	305	350	390	465	505	540	610
Sand and gravel	125	155	185	210	235	255	295	340	380	455	495	530	600
Good, common earth	105	135	165	190	210	230	265	305	340	375	410	445	510
Clay, hard and tough	90	110	135	160	180	195	230	270	305	340	375	410	475
Clay, wet and sticky	55	75	95	110	130	145	175	210	240	270	300	330	385

Note: These data assume average job conditions, a 90-degree swing, optimum depth of cut, and material loaded into hauling units.

4.2.7   Trucks

When large amounts of material are to be hauled over public roads or when prolonged grades are involved, test runs are made with loaded and empty trucks of the same type that will be used on the proposed work to determine the average travel speeds. Typically, trucks will travel at average speeds of 40 miles per hour on good county roads and 26 miles per hour in urban areas where traffic is light.

Although on-highway trucks may reach a top speed of 40 miles per hour on well-maintained, dirt haul roads, the average speed, in most cases, will be about 27 miles per hour.

Whenever trucks are used, a load counter should keep a log of the truck cycle times; when large amounts of material are being hauled from off-job locations, a truck foreman should be used to keep the operation running efficiently.

Good judgment is required when hauls must be made across railroad tracks. In all cases, data on the anticipated train traffic during the haul period must be obtained and carefully considered. When large quantities of material must cross heavily traveled tracks, temporary haul bridges or conveyor belt systems may be necessary.

4.2.8   Draglines*

Dragline production (Table 4.1) may be roughly estimated from the data that follow, which assume that mass excavation is being performed, close adherence to lines and grade is not required, and the boom length is at or near the lower end of the ranges shown in Table 4.2.

Assume the following conditions:

$$\begin{array}{ll} \text{Bucket size} & \text{1 CY} \\ \text{Swing} & \text{120 degrees} \end{array}$$

TABLE 4.2   Standard Boom Lengths for Draglines

Bucket size (CY)	$\frac{3}{4}$	1	$1\frac{1}{2}$	2	3	4	5
Boom length (feet)	35–50	40–55	50–70	50–90	60–110	70–120	70–130

*Tables 4.1 through 4.4 are copyrighted material taken from Technical Bulletin No. 4 of the Power Crane and Shovel Association and are reproduced here with their permission.

Depth of cut   8 feet
Material   Sand
Working hour   50 minutes

The optimum depth of cut, from Table 4.3, is 6.6 feet, and the job cut is

$$\frac{8}{6.6} \times 100 = 121 \text{ percent of optimum}$$

From Table 4.1, the estimated production rate is 155 BCY per 60-minute hour, and from Table 4.4, the correction factor is 0.90. The effective production rate is estimated to be

$$155 \times 0.90 \times \frac{50}{60} = 116 \text{ BCY per hour}$$

Cross Reference: Wiley Series of Practical Construction Guides, *Construction Guide for Soils and Foundations* by Robert G. Ahivin and Vernon A. Smoots.

TABLE 4.3   Optimum Depths of Cut in Feet for Short-Boom Dragline Performance

	Bucket Capacity (CY)												
Class of Material	$\frac{3}{8}$	$\frac{1}{2}$	$\frac{3}{4}$	1	$1\frac{1}{4}$	$1\frac{1}{2}$	$1\frac{3}{4}$	2	$2\frac{1}{2}$	3	$3\frac{1}{2}$	4	5
Light moist clay or loam	5.0	5.5	6.0	6.6	7.0	7.4	7.7	8.0	8.5	9.0	9.5	10.0	11.0
Sand or gravel	5.0	5.5	6.0	6.6	7.0	7.4	7.7	8.0	8.5	9.0	9.5	10.0	11.0
Good common earth	6.0	6.7	7.4	8.0	8.5	9.0	9.5	9.9	10.5	11.0	11.5	12.0	13.0
Clay, hard and tough	7.3	8.0	8.7	9.3	10.0	10.7	11.3	11.8	12.3	12.8	13.3	13.8	14.3
Clay, wet and sticky	7.3	8.0	8.7	9.3	10.0	10.7	11.3	11.8	12.3	12.8	13.3	13.8	14.3

TABLE 4.4   Correction Factors for the Effects of Depth of Cut and Angle of Swing on Dragline Production

Depth of Cut (% of optimum)	Angle of Swing (in degrees)							
	30	45	60	75	90	120	150	180
20	1.06	0.99	0.94	0.90	0.87	0.81	0.75	0.70
40	1.17	1.08	1.02	0.97	0.93	0.85	0.78	0.72
60	1.25	1.13	1.06	1.01	0.97	0.88	0.80	0.74
80	1.29	1.17	1.09	1.04	0.99	0.90	0.82	0.76
100	1.32	1.19	1.11	1.05	1.00	0.91	0.83	0.77
120	1.29	1.17	1.09	1.03	0.98	0.90	0.82	0.76
140	1.25	1.14	1.06	1.00	0.96	0.88	0.81	0.75
160	1.20	1.10	1.02	0.97	0.93	0.85	0.79	0.73
180	1.15	1.05	0.98	0.94	0.90	0.82	0.76	0.71
200	1.10	1.00	0.94	0.90	0.87	0.79	0.73	0.69

CHAPTER 5

Structure Excavation and Structure Backfill

Structure excavation and structure backfill are taken off together but are normally bid as two separate job items. The quantities involved are relatively small; the upper vertical pay limit for structure excavation is frequently the lower pay limit for a mass excavation item, and the upper limit for the structure backfill may be the subgrade for an embankment.

5.1 STRUCTURE EXCAVATION

Typically, the excavated material is stockpiled at the site and is subsequently used for structure backfill. After the backfill is completed, any excess material is incorporated into an adjacent fill at no additional cost or is spread out and "lost" during the final cleanup operation.

Structure excavation is normally done with equipment to within about ±0.1 foot of finish grade, and the fine grade is made by hand. When concrete structures are involved, the area to be fine graded is taken off and priced with the concrete; typical costs for such fine grading are shown in the unit cost tables in Chapter 8. The fine grading costs relating to pipe excavation are included in the unit costs for laying the pipe.

5.1.1 Hand Excavation

A detailed analysis of hand excavation costs is seldom warranted since the total cost for hand excavation normally represents an insignificant portion of the overall job cost.

In sandy soil free from rocks where the material can be cast to the side, one laborer will excavate about 1 CY per hour. If picks and wheelbarrows are required, this basic cost increases.

When deep underpinning pits are excavated by hand, a power winch and buckets are required. A three-man crew is used with one man in the pit, one man at the head of the pit, and one man on the winch. In easy digging such a crew will excavate from 0.4 to 0.6 CY per hour when working in a pit 4 feet square.

5.1.2 Unit Costs

Unit costs for structure excavation are shown in Table 5.1. The unit costs for labor include the cost of a grade checker as well as the costs for that incidental hand excavation required to remove material from the bottom corners of the excavated sections.

The horizontal pay limits for pipe excavation are normally specified to be vertical planes 1 foot outside the neat lines of the pipe. In most cases trenches less than 5 feet deep are dug with vertical banks, and the volume of the excavation required will equal that shown on the engineer's estimate.

When estimating pipe excavation costs, the trench depths must be determined since vertical, unshored banks are limited to 5 feet in height by safety regulations. On many jobs, profiles for the pipe work are not shown, but the engineer's estimate for the pipe excavation is tabulated for each run of pipe. In such cases the average trench depth is determined for each run of pipe as follows:

$$\bar{D} = \frac{27\,E}{LW}$$

TABLE 5.1 Unit Costs for Structure Excavation

Type of Structure or Activity	Labor		Rental		Unit of Measure
	Unit Cost	Index	Unit Cost	Index	
Box culverts[a]	$0.90	$K_L = 13.48$	$2.65	$K_{Cr} = 75$	CY
Flat slab bridges[b]	0.70	$K_L = 13.48$	1.75	$K_{Cr} = 75$	CY
Girder bridges	0.80	$K_L = 13.48$	3.10	$K_{Cr} = 75$	CY
Pipe excavation[c]	1.20	$K_L = 13.48$	2.00	$K_{BH} = 25$	CY
Grade beam excavation	1.10	$K_L = 13.48$	1.70	$K_{BH} = 25$	CY

[a]These costs apply to box culverts located in existing channels where the bulk of the excavation is for the wingwalls.
[b]Similar to bridge shown on Figure 8.11.
[c]Depth about 5 feet and work area relatively flat.

Note: These unit costs are applicable to work in ordinary earth and include the costs for checking grades. The unit costs should be applied to the true quantity of excavation, which may be greater than that shown on the engineer's estimate. On-and-off charges for equipment are additional.

where

$\bar{D}$ = average depth of trench in feet

E = the engineer's estimated cubic yards for the run of pipe being considered

L = the length of the run in feet

W = the width between the horizontal pay limits in feet

Typically, on freeway work, the average depth so determined will closely approximate the true average depth, and this can be verified, when necessary, by a review of the typical cross sections for the area under consideration.

When trenches are to be between about 5 and 8 feet in depth, the upper layer of material may be moved to one side with a bulldozer, and the bottom 5 feet excavated with vertical banks. During the backfill operation, the upper layer is replaced and compacted by the mass backfill method.

The unit costs for equipment shown in Table 5.1 reflect the low production rates achieved in this type of excavation. Thus the effective production rate for box culvert excavation is $75/2.65 = 28$ CY per hour and is based on using a $\frac{3}{4}$-CY dragline for the box section and a $\frac{3}{4}$-CY clamshell bucket for the headwalls. It is assumed the box culvert is in a natural channel or in a channel excavated under a mass excavation item and that only a small amount of excavation is required for the box section.

The low production rates for structure excavation result from the small amounts of excavation possible per setup, the necessity to closely control the lines and grade, and the inability of the equipment to set and bail.

The unit costs for girder bridges include the costs of excavating for the abutments and for the piers, and all unit costs shown assume that the excavated material is stockpiled at the site.

Well-defined hand excavation such as that required for the retaining wall foundation keys, illustrated in Chapter 8, is taken off and priced with the related concrete. Incidental amounts of hand excavation are, however, required in connection with excavation done with machines.

When unit costs for structure excavation are derived from production rates for equipment, the labor cost for the grade checker is easily determined and is applied to the total amount of machine excavation. The cost of the incidental hand excavation is, in such cases, based on a careful consideration of the typical excavation cross sections.

Thus, if a total of 1000 CY of structure excavation has been taken off, and if the estimator concludes that 5 percent of this material must be removed by hand, the unit cost for equipment would be applied to 950 CY, the unit cost for hand excavation would be applied to 50 CY, and the total costs for these two items would be entered in the rental and labor columns of the estimate sheet.

5.1.3 Footing Excavation with Trenching Machines

Wheel-type trenching machines are frequently used to excavate for the grade beams in building foundation work. The grade beams are then poured neat, the structure backfill is eliminated, and savings are achieved.

When trench lengths are greater than 100 feet, trenches 20 inches wide by 3 feet deep can be dug in ordinary earth at a rate of about 280 linear feet (LF) per hour. Short trenches averaging 25 feet long can be dug at a rate of about 100 LF per hour. In all cases an allowance must be made for moving the machine on and off the job, and a grade checker will be required.

5.1.4 Sheeting and Shoring

Current safety regulations require trenches or banks greater than 5 feet in depth or height to be supported with sheeting or laid back to a safe slope which must be no steeper than $\frac{3}{4}$ horizontal to 1 vertical.

Where large areas are to be excavated and space permits, it is usually cheaper to slope the banks than to shore them. When excavating for retaining walls or for building foundations and basements, however, sophisticated shoring systems are frequently required to support the banks and adjacent roads or buildings. Such work is normally done on a subcontract basis by pile driving or foundation contractors specializing in this type of construction.

On occasion, for temporary support, small amounts of short steel sheet piles may be set and driven by the general contractor using a truck crane equipped with hanging leads and a drop hammer. A temporary waling system is required and it may consist of a 4 × 12 sill at the bottom, 4 × 12 posts 5 feet high at 6 foot centers, and a 4 × 12 cap. The waling is braced with 2 × 8's and 2 × 4 stakes at 6-foot centers.

All the piles are set and braced against the wales before starting to drive. The entire row is then driven in a series of passes, and the driving depth is about 12 inches per pass. An inexperienced three-man crew will set and drive sheet piles 12 feet long in this manner at a rate of about 90 square feet per hour. Based on K_{6+2} = 1000, the wales will be set and stripped for about $450 per MFBM. In addition to the on-and-off charges for the crane and leads, about 6 hours will be required to rig up and dismantle the leads and hammer.

A pit, located in sand, 6 × 26 feet in plan and 12 feet deep was sheeted with 3 × 12 vertical timbers supported by three sets of 6 × 6 wales with 6 × 6 struts. The timber was set for $150 per MFBM based on K_{6+2} = 1000. The excavation was made with a $\frac{3}{4}$-CY clamshell bucket at an effective rate of 8 CY per hour.

Banks 30 feet high were shored by using properly reinforced cast-in-drilled-hole (CIDH) piles together with 2 × 12 horizontal sheeting. The 16-inch-diameter piles were installed in a row on 3-foot centers. The pile tips extended 12 feet below the bottom of the proposed excavation, and the material was moist sand.

The excavation along the face of the piles was done with a $\frac{3}{4}$-CY clamshell bucket, and pieces of 2×12 sheeting approximately 2 feet long were placed horizontally, between and slightly behind the centerline of the piles, as the excavation progressed.

The excavation was done at an effective rate of 20 CY per hour, and the 2×12 material was installed for $1730 per MFBM based on $K_L = 13.48$. The piles were supported by two steel wales together with timber struts.

5.2 STRUCTURE BACKFILL

Structure backfill on first-class work must be compacted to a relative compaction of 90 or 95 percent and is a costly item. Unfortunately, some smaller agencies specify such compaction but do not take tests or require it in the field. Small jobs have been lost when the contractor bid the price to place structure backfill as specified on projects where the competition knew the specified compaction would not be required.

The outside horizontal pay limits for structure backfill are the same as those for structure excavation, and the true volume placed will often amount to two or three times the amount shown on the engineer's estimate.

Public agencies frequently require sand to be used for backfill material within the pay limits in those locations where relative compaction of 95 percent is required. Such sand will normally have a specified gradation and sand equivalent. The sand may be available on the job site, but in many cases it must be ordered from an outside source, and from 1.7 to 1.9 tons will be required per compacted cubic yard.

The horizontal pay limit for structure backfill is often a vertical plane 1 foot away from a concrete wall. Typically, the true section to be backfilled will be 1 foot wide at the bottom bounded by the wall on one side and a sloping bank on the other.

When imported sand is required in such a case, sufficient sand must be ordered to place a vertical sand section at least 1.5 feet wide against the concrete since it is not practical to place the sand to an exact line. The imported sand and the material from the excavation are placed in layers 6 to 8 inches thick and as near as possible to their required locations by a wheel loader during the compaction operation.

5.2.1 Unit Costs

Typical unit costs for structure backfill in confined areas are shown in Table 5.2. As used here, a confined area is one that is too small to permit the use of road-type compaction equipment.

Except for the pipe backfill, the compaction is assumed to be done with a small loader-backhoe equipped with a hydraulic backfill compactor. Hand-operated backfill compactors are assumed for the pipe backfill since the backhoe-mounted compactors will damage or distort pipes.

The unit costs for labor and equipment include the costs for placing imported sand, when required, as well as the costs for applying the water needed for compac-

TABLE 5.2 Unit Costs for Structure Backfill

Type of Structure or Activity (Relative Compaction)	Labor		Rental		Unit of Measure
	Unit Cost	Index	Unit Cost	Index	
Box culverts (90–95%)	$2.50	$K_L = 13.48$	$3.20	$K_{BH+C} = 31^a$	CY
Box culverts (no testing)	0.80	$K_L = 13.48$	1.25	$K_{BH+C} = 31$	CY
Flat slab bridges (95%)	3.25	$K_L = 13.48$	4.30	$K_{BH+C} = 31$	CY
Girder bridges (95%)	2.30	$K_L = 13.48$	3.60	$K_{BH+C} = 31$	CY
Pipe backfill (90–95%)	6.80	$K_L = 13.48$	4.70	$K_{BH+T} = 30^b$	CY

$^a K_{BH+C}$ = hourly rental rate for a $\frac{1}{2}$-CY wheel loader-backhoe equipped with a hydraulic backfill compactor.
$^b K_{BH+T}$ = the sum of the hourly rental rates for a $\frac{1}{2}$-CY wheel loader-backhoe and a hand-operated backfill compactor.

Note: The unit costs shown are for work in confined areas and include the labor cost to apply water. The costs to furnish the water required for compaction are additional. The unit costs should be applied to the true quantity of backfill, which may be greater than the amount shown on the engineer's estimate. On-and-off charges for equipment are additional.

tion. The material cost for the imported sand and the cost to supply the water to the site are additional in all cases.

Sandy material may be flooded and compacted with concrete vibrators when job conditions permit excess water to drain without damage to the work. Material was placed in a 6-foot-wide area with a $\frac{3}{4}$-CY clamshell bucket at an effective rate of 17.5 CY per hour. The material was compacted with two two-man crews using one 3-HP vibrator per crew. When using concrete vibrators in sand, the vibrator heads must be water cooled when working. This is achieved by furnishing a constant supply of water to each head with a garden hose.

5.2.2 Mass Structure Backfill

Where the available space permits and where large amounts of structure backfill are involved, mass structure backfill is placed with a road-type compacting equipment. In the typical case, however, the job conditions relating to the backfill operations do not permit maximum production to be achieved.

5.2.3 Combined Structure Backfill

Maximum economy may require the structure backfill for some structures to be placed by a combination of methods. For example, assume the back face of the retaining wall shown in Figure 5.1 is to be backfilled to line *EF*. Assume further that material for the backfill is stockpiled about 100 feet to the right of point *F*, that the wall section shown has an effective length of 200 feet, and that access is available at the ends of the section.

The area to be backfilled is divided into three zones as shown. Zone 1 will be placed with a loader-backhoe equipped with a hydraulic backfill compactor. Zone 2 will be placed with a double 5 × 5 sheepsfoot roller, and zone 3 will be placed

Figure 5.1. Combined structure backfill.

by hand, concurrently with the material in zone 2. The hand backfill in zone 3 is necessary because the sheepsfoot roller cannot work within 1 to 1.5 feet of the wall.

Zone 3 is drawn as shown to simplify the takeoff, although this material will, in fact, be placed against the battered back face of the wall.

The lower limit of zone 2 is determined as follows: line *CD* is located so that its length is 12 feet, which is a minimum width for the sheepsfoot roller. Line *AB* is arbitrarily set 2 feet below line *CD* to increase the area of zone 2 as much as possible, and the triangular cut at the bank will be made and backfilled at no additional cost during the mass backfill operation.

The backfill is taken off as follows (SF = square feet):

Zone 1:

$$\text{Area} \quad \frac{10 + 6.5}{2} \times 3.5 - 1.3 \times 5.5 = 21.7 \text{ SF}$$

$$\text{Volume} \quad 21.7 \times \frac{200}{27} = 161 \text{ CY}$$

Zones 2 + 3:

$$\text{Area} \quad \frac{27.8 + 10}{2} \times 17.8 = 336.4 \text{ SF}$$

Zone 3:

$$\text{Area} \quad 1.5 \times 17.8 = 26.7 \text{ SF}$$

Volume $$26.7 \times \frac{200}{27} = 198 \text{ CY}$$

Zone 2:

Area $$336.4 - 26.7 = 309.7 \text{ SF}$$

Volume $$310 \times \frac{200}{27} = 2300 \text{ CY}$$

The additional volume of backfill due to the battered back face is $\frac{1}{2} \times 0.8 \times 20 \times 200/27 = 59$ CY, and this volume is added to the volume calculated for zone 2. The error that results from neglecting zone 1 in this procedure is negligible, and the final backfill volumes are taken as follows:

Zone 1	161 CY
Zone 2	2360 CY
Zone 3	198 CY
Total	2719 CY

Assuming the horizontal pay limits for structure backfill are vertical planes 1 foot outside the neat concrete lines, the engineer's estimate for this portion of backfill would be

$$2719 - \frac{21.3 \times 21.3 \times 200}{2 \times 27} = 1039 \text{ CY}$$

Since the hand backfill in zone 3 must be placed concurrently with the mass backfill in zone 2, it is now necessary to arrive at reasonable, balanced production rates for these two items.

A three-man crew supplied with material will compact it at an effective rate of about 6 CY per hour. This assumes one man operating the compactor, which may be a vibrating plate, a vibrating roller, or a rammer-type unit, one man applying water with a garden hose, and one man shoveling and leveling ahead of the compactor.

Thus, two crews will compact the material in zone 3 in $\frac{198}{12} = 17$ hours while the sheepsfoot will be compacting the material in zone 2 at an average rate of $\frac{2360}{17} = 139$ compacted CY per hour.

This is a slow production rate for the roller, and we note that it could be doubled if more than two hand backfill crews were used. Considering the cramped work area and the relatively short duration of the backfill operation, however, the 139-CY-per-hour rate is selected as the fastest practical rate in this particular instance.

The backfill material for zones 2 and 3 will be pushed over the bank by a bulldozer working at the top, and one 185-HP bulldozer will be adequate.

The unit costs for placing backfill in zones 1, 2, and 3 are determined as follows:

Zone 1. The backfill in zone 1 is similar to that required for bridge pier footings except that the backfill material is further away. The unit costs for girder bridge backfill shown in Table 5.2 will be used, but the rental cost will be increased by an arbitrary $1.00 per CY to cover the cost of moving the material into the hole. The $1.00 per CY allowance assumes that a bulldozer or perhaps a wheel loader will be available somewhere on the job at the time this work is done. Thus the unit costs for the zone 1 backfill will be:

Labor $2.30 per compacted CY
Rental $4.60 per compacted CY

Zone 2. The material will be furnished by a 185-HP bulldozer, spread by a motor grader, and compacted with a 185-HP bulldozer pulling a sheepsfoot roller, and water will be supplied from a water truck. A grading foreman will be required, and his hourly wage rate, including all fringes, is assumed to be $18.50 per hour. The assumed rental rates for the equipment are as follows:

185-HP bulldozer	$ 50 per hour
185-HP bulldozer	50
Double 5 × 5 sheepsfoot roller	5
125-HP blade	39
Water truck	30
Total hourly cost	$174

The unit costs for the zone 2 backfill will be:

$$\text{Labor} \quad \frac{18.50}{139} = \$0.14 \text{ per compacted CY}$$

$$\text{Rental} \quad \frac{174}{139} = \$1.25 \text{ per compacted CY}$$

Zone 3. Assuming the hourly rental rate for the hand-operated backfill compactors to be $5.00 per hour and $K_L = 13.48$, the unit costs for the hand backfill in zone 1 will be:

$$\text{Labor} \quad \frac{6 \times 13.48 \times 17}{198} = \$6.94 \text{ per compacted CY}$$

$$\text{Rental} \quad \frac{2 \times 5 \times 17}{198} = \$0.86 \text{ per compacted CY}$$

The preceding unit cost for rental does not include any costs for furnishing the backfill material since this material will be furnished and spread by the mass back-fill equipment.

The time of completion estimate for backfilling the wall is made using the production rates discussed in the foregoing together with the production rate for the backfill in zone 1, which, from Table 5.2, is $31/3.60 = 8.6$ CY per hour.

The on-and-off charges for the equipment must be added. Assume the equipment will be moved 4 miles through suburban streets and 11 miles on good country roads. Assume also that the hourly rental rate for a flatbed truck is $30 per hour. Driving time for the flatbed is estimated as $\frac{8}{25} + \frac{22}{40} = 0.9$ hour round trip. Allowing 1 hour to load and unload the equipment, the total on-and-off time per piece is $1.9 \times 2 = 3.8$ hours, which is rounded off to 4 hours.

The two bulldozers, the sheepsfoot roller, and the blade will each cost $4 \times 30 = \$120$ to move on and off. The water truck will cost 1 hour's rental, as will the loader-backhoe, which arrives on its own trailer. Thus, the total on-and-off charges will be:

4 pieces at $120	$480
Water truck, 1 hour	30
Loader-backhoe, 1 hour	31
Total	$541

An allowance should be made on the estimate for the cost of the garden hoses used to supply water to the hand backfill operation and, if necessary, the costs for a small (250-gallon) water tank together with a 2-inch centrifugal pump. It is assumed that the water supply for the water truck is developed under a separate job item.

Note: Labor, rental, and unit costs are used for illustration purposes only and do not represent present-day costs.

Roadway Construction

6.1 ROADWAY EXCAVATION AND IMPORTED BORROW

The job items for roadway excavation and imported borrow represent similar types of work, and on many projects, the combined costs for these two items will constitute more than 50 percent of the total job cost. Roadway excavation is made with off-highway equipment, wheel tractor-scrapers are commonly used to haul the material, and the haul lengths are dictated by the road profiles.

The type of equipment used to haul imported borrow will depend on whether the hauls are on public roads or off-highway and on the lengths of hauls required. The haul lengths will depend on the location of the borrow pit with respect to the related embankments; it is imperative that these haul lengths be kept as short as possible.

Roadway excavation is the excavation required to grade and construct the roadway. The price paid for roadway excavation includes the costs of excavating, sloping, shaping, loading, hauling, spreading, and compacting the material and the cost of making the grade for the roadway. Roadway excavation is paid for in the cut, and the unit of measure is bank yards.

When the excavated material must be hauled more than 10 stations (1000 feet), a separate item for overhaul may be included in the engineer's estimate.

6.1.1 The Roadway Excavation Takeoff

The takeoff will include the following items:

1. A check of the engineer's estimated yardage for the item.
2. A takeoff of any excavation or work required by the specifications which is not included in the engineer's estimate but whose cost is specified to be included in the price bid for roadway excavation.

3. A takeoff of that excavation which is neither specified nor included in the engineer's estimate but which must be done to complete the roadway excavation item.
4. A takeoff of the areas on which a grade must be made.
5. A determination of the average haul distances for the various sections of roadway excavation.
6. An estimate of the number of flares, barricades, signs, and flagmen required for traffic control and for the protection of the public during the course of the roadway excavation work.

6.1.1.1 CHECKING THE ENGINEER'S ESTIMATE. The quantities of excavation and embankment are tabulated on the road profiles. On small jobs they are tabulated per station, and on large jobs they may be tabulated for each ten stations.

The sum of the tabulated excavation quantities should equal the engineer's estimate for the item, but in many cases the estimate will have been arbitrarily increased by about 5 percent for budgetary reasons.

The tabulated quantities of excavation and embankment will be true quantities. To determine haul distances, one or the other of these quantities must be adjusted by the grading factor.

6.1.1.2 SPECIFIED WORK NOT INCLUDED IN THE ENGINEER'S ESTIMATE. The specifications for roadway excavation frequently require performing a number of related work items whose costs are specified as included in the price bid for roadway excavation. Such work may include certain clearing and grubbing operations, saw cutting of pavements, compacting original ground, undercutting and recompacting material in certain areas, or the benching of hillsides against which embankments will be constructed. The additional excavation required because of shrinkage resulting from such compacting or recompacting work will not be included in the excavation quantities tabulated on the profiles.

Because of the inherent imprecision in determining the grading factor and because of the additional compaction discussed in the preceding, some roadway excavation jobs will end up short of dirt. In such cases the costs of obtaining and placing the additional material required to complete the embankments are normally covered under a contract change order.

When embankments are constructed on hillsides or against existing embankments, the existing slopes must be cut unto, horizontally, for a specified distance, and the work is brought up in layers. The material thus cut is recompacted with the embankment material as the work progresses, and the excavation required is not included in the engineer's estimate.

The takeoff for such bench excavation is based on assumed typical sections, the excavation is made with a bulldozer, and no other costs need be considered.

6.1.1.3 ADDITIONAL EXCAVATION REQUIRED FOR CONSTRUCTION. A typical cross section for a freeway ramp constructed of embankment is shown in Figure 6.1 together with a sketch used to determine the additional excavation required to construct the shoulders.

The hinge points, designated HP, are located at the intersections of the upper surface of the roadway and the side slopes. The area between the hinge points is the roadbed, and area *ABCD* is the structural section, which, in this case, consists of 0.5 feet of asphaltic concrete on 1.05 feet of aggregate base.

The embankment section must be completed before placing materials in the structural section. To complete the embankment, a grade must be made on the shoulders and on the subgrade, which is line *BC* on the cross section.

Since the 5-foot shoulder on the left and the 3-foot shoulder on the right are both too narrow to permit road-building equipment to operate, enough additional embankment to furnish a minimum total top width of 12 feet must be placed at each shoulder. The additional excavation required to place and remove this temporary embankment material will not be included in the engineer's estimate, and thus no payment will be made for doing this work.

Referring to the sketch on Figure 6.1, areas *ABFE* and *CDJI* represent additional embankment required above the subgrade:

$$\text{Area } ABFE = 14.7 \text{ SF}$$
$$\text{Area } CDJI = 15.2 \text{ SF}$$

The triangular area at the top of *ABFE* results from the 7 percent difference in the slopes between the shoulder and the structural section. A similar situation near point *J* is neglected since the difference in slopes is only 3 percent and the length involved is only 1 foot.

Assuming the length of the ramp to be 700 feet, the total volume of additional material above the subgrade is

$$700 \times \frac{14.7 + 15.2}{27} = 775 \text{ CY}$$

Figure 6.1. Road cross section.

After the shoulders have been completed, the additional material must be re-moved and placed in an embankment. On wide structural sections where the dis-tance *FI* is about 14 feet or more, the subgrade is left low enough, in this area, to receive the additional material, which is excavated and moved to its final position with a motor grader. When structural sections are too narrow to permit this proce-dure, the additional material is excavated, hauled, and placed into a nearby embank-ment.

In both procedures noted in the foregoing, one additional roadway excavation operation is required for the material used to temporarily widen the shoulder sec-tions, which, in this case, amounts to 775 CY. The cost to excavate this material is based on the unit costs for roadway excavation made within the free-haul limit.

Sometimes the structural sections are narrow, there are no nearby embankments, and the additional material must be placed below the subgrade of the structural section in which it is used. This procedure requires additional unpaid-for excava-tion, and it will be necessary to widen the shoulder sections one side at a time.

In such cases, the embankment section is constructed to line *KL,* which is paral-lel to and below the subgrade. Since area *CDJI* is greater than area *ABFE,* the right shoulder is completed first, which requires area *DJHL* to be placed.

Upon completion of the right shoulder, area *DJHL* is removed and placed in the left-shoulder area *AKGE;* the excess material is used in the shoulder section to the left of line *AK.*

After completion of the left shoulder, area *ABFE* is removed and placed in area *CFGL,* which requires that the distance *BK* be selected so that areas *ABFE* and *CFGL* are equal. Assuming distance *BK* to be about 1 foot, the average width of area $CFGL = 22 - 9.1 - 0.5 = 12.4$ feet, and we have

$$BK \ (12.4) \ = \ 14.7$$

or

$$BK \ = \ 1.2 \text{ feet}$$

The areas for the shoulder sections shown in Figure 6.1 are calculated and listed as follows:

$$
\begin{array}{ll}
\text{Area } DJHL & = \ 28.6 \text{ SF} \\
\text{Effective area } DJHL & = \ 28.6/0.85 = 33.6 \text{ SF} \\
\text{Area } AKGE & = \ 26.4 \text{ SF} \\
\text{Area } CBKL & = \ 26.4 \text{ SF} \\
\\
\text{Effective area } CBKL & = \ 26.4/0.85 = 31.1 \text{ SF}
\end{array}
$$

An assumed grading factor is applied to areas *DJHL* and *CBKL* since both will, in effect, be filled with material from the cut. The additional, unpaid excavation is determined as follows:

+ 33.6 SF Excavate and place in right shoulder.

+ 28.6 SF Excavate from right shoulder and place in left shoulder.

− 2.2 SF Area *DJHL* — area *AKGE*. This represents excess material placed to the left of line *AK*. This material is paid for.

+ 14.7 SF Excavate area *ABFE* and place in area *CFGL*.

− 31.1 SF Effective area *CBKL* is paid for under the roadway excavation item.

    43.6 SF Additional, unpaid excavation area.

Assuming the ramp to be 700 feet long, the additional unpaid excavation is

$$43.6 \times \tfrac{700}{27} = 1130 \text{ CY}$$

To save time, the estimator calculates this yardage by more approximate methods. The triangular area at the top will be neglected, and the remaining areas are assumed to be rectangular in shape. Since typical jobs may have twenty to forty ramps, a significant amount of additional excavation may be involved, and its additional cost should not be overlooked.

6.1.1.4 AREA TAKEOFF FOR MAKING GRADE. The final operation for the roadway excavation item is making the grade for the roadbed. To determine the required areas, widths are taken from the typical road cross sections; these are the total widths between hinge points. The corresponding lengths are taken from the road plans, and the resulting areas are expressed in square yards.

Referring to Figure 6.1 and assuming that R5 is 700 feet long, the area for making the grade on this ramp would be $30 \times 700 \times \tfrac{1}{9} = 2330$ SY (square yards). When the unit costs for making the grade on roadbeds are applied to the total of all such areas, the resulting costs will include the costs for making the grade on the shoulders as well as on the subgrade of the structural sections.

It is not necessary to take off the areas of the cut or embankment slopes. The allowable grade tolerance is normally ±0.5 foot for cut slopes and from ±0.5 to ±1.0 foot for embankment slopes. With proper care, these tolerances can be met fairly well as the work is done, but the embankment slopes may require trimming, under the item for finishing roadway.

The median is the portion of a divided highway separating the traveled ways and includes the inside shoulders. In taking off the median areas for grade-making purposes, however, the widths are taken between the inside shoulders. The median areas are kept separate from the roadbed areas since different unit costs apply.

Many jobs have items for both roadway excavation and imported borrow. In such cases the grade-making areas for the entire job are taken off in one operation, and no attempt is made to separate the embankment sections made under either item. The total areas so determined may be somewhat arbitrarily apportioned between the two items or the entire grade-making operation may be charged to the larger item.

6.1.1.5 AVERAGE HAUL DISTANCES. The determination of the average haul distances relating to the different segments of excavation is an important part of the roadway excavation takeoff since haul lengths directly affect the cycle times for the hauling equipment.

The determination of haul distances requires that certain cut volumes be balanced with certain fill volumes. To achieve such an earthwork balance, either the cut or the fill volumes, tabulated on the road profiles, must be adjusted by the grading factor.

6.1.2 Haul Distances by Arithmetic Methods

On small jobs, haul distances may be determined by inspection or arithmetically, as shown in Table 6.1. This table assumes that excavation and embankment quantities are tabulated on the road profiles in two-station intervals and that the grading factor is 0.85.

The quantities in columns 2 and 4 are taken directly from an assumed road profile, the embankment quantities being considered negative and the excavation quantities being considered positive. The grading factor is applied to the embankment quantities, and the adjusted embankment quantities are listed in column 3. Column 5 contains the net cubic yards per interval and is the algebraic sum of the quantities in columns 3 and 4. The center of mass (actually, the center of volume) for each quantity, listed in column 5, is assumed to be at the center of each two-station interval.

The net volumes in column 5 are balanced in column 6, starting at the grade point, station 11, where the job passes from cut to fill. The 53 CY of fill required at station 11 is made from the excavation at station 9. This leaves 747 CY of excess excavation at station 9, which is taken to station 13. This leaves 194 CY of fill still

TABLE 6.1 Haul Distances: Arithmetic Method

(1) Station	(2) Embankment C	(3) Embankment 0.85 CY	(4) Excavation CY	(5) Net CY	(6) Earthwork Balance	
0						
2	− 200	− 235	+ 400	+ 165	− 165	= 0
4	—	—	+ 1000	+ 1000	− 12	= + 988
6	—	—	+ 1200	+ 1200	− 271	= + 929
8	—	—	+ 1100	+ 1100	− 194	= + 906
10	—	—	+ 800	+ 800	− 53	= + 747
12	− 300	− 353	+ 300	− 53	+ 53	= 0
14	− 800	− 941	—	− 941	+ 747	= − 194
16	− 1000	− 1177	—	− 1177	+ 906	= − 271
18	− 800	− 941	—	− 941	+ 929	= − 12
20	− 1100	− 1294	—	− 1294	+ 988 + 165	= − 141
22	− 200	− 235	+ 400	+ 165	− 141	= + 24
Total	− 4400	− 5176	+ 5200	+ 24		

required at station 13, which is taken from the excavation at station 7. This process is continued, and finally, at station 21, we have an excess of 24 CY of cut, which checks the column 5 total.

The haul distances for the excavation quantities are taken from column 6, where each of the excavation quantities appears both at its station of origin and at the station to which it is hauled. The distances and quantities are tabulated as follows:

Stations		Haul Distance	
From	To	(Stations)	CY
9	11	2	53
9	13	4	747
7	13	6	194
7	15	8	906
5	15	10	271
5	17	12	929
3	17	14	12
3	19	16	988
1	19	18	165
19	21	2	141
			Total CY hauled 4406

The total CY hauled does not equal the total of the cubic yards of excavation tabulated on the profile and listed in column 4 of Table 6.1. This results from the use of net quantities for those segments consisting of both cut and fill. Referring to Table 6.1, we have 235 CY of excavation at station 1300 CY at station 11, and 235 CY at station 21, or a total of 770 CY of excavation, which is not included in the haul yardage. As discussed in the following section, this is called sidehill excavation, and together with the 24 CY of waste, it will appear on the engineer's estimate and will be paid for. The sum of the haul yardage, the sidehill excavation, and the waste equals 5200 CY, which checks the total for column 4 in Table 6.1.

The average haul distance from the center of mass of the excavation to the center of mass of the embankment for the material tabulated in the preceding is (2 × 53 + 4 × 747 + 6 × 194 + 8 × 906 + 10 × 271 + 12 × 929 + 14 × 12 + 16 × 988 + 18 × 165 + 2 × 141) ÷ 4406 = 10 stations. In addition, the 794 CY of sidehill excavation and waste will be moved about one station.

For estimating purposes, however, the preceding haul yardage would be broken down into yards hauled ten stations or less and yards hauled more than ten stations. For these purposes, the preceding tabulation is restated as follows:

Haul Distance (stations)	CY
2	53
4	747

6	194
8	906
10	271
2	141
Total yards (hauled ten stations or less)	2312

and

Haul Distance (stations)	CY
12	929
14	12
16	988
18	165
Total yards (hauled more than ten stations)	2094

Calculating the average haul distances for these yardages, as before, we get

2312 CY hauled 6 stations

and

2094 CY hauled 14 stations

In addition, the 794 CY of sidehill excavation and waste will be moved about one station.

The ten station average haul for the entire section can also be determined without resorting to the arithmetical balancing procedure shown in Table 6.1, column 6. To do this, the centers of mass must be located for the section in cut and for the section in fill. Taking the data from Table 6.1, we have:

Station	Net CY Excavation
1	165
3	1000
5	1200
7	1100
9	800
Total	4265 CY

Station	Net CY Embankment
11	53
13	941
15	1177
17	941
19	1294
Total	4406 CY

Taking moments about station 0, the station of the center of mass for the excavation section is $(1 \times 165 + 3 \times 1000 + 5 \times 1200 + 7 \times 1100 + 9 \times 800) \div 4265 = 5.6$. Again, taking moments about station 0, the station of the center of mass of the embankment section is found to be station 16.1.

Thus we have

$$4265 \text{ CY haul } (16.1 - 5.6) = 10.5 \text{ stations}$$

and

$$\frac{141}{4406} \begin{array}{l} \text{CY haul } (21 - 16.1) = 4.9 \text{ stations} \\ \text{CY total} \end{array}$$

The average haul for this yardage is $(4265 \times 10.5 + 141 \times 4.9) \div 4406 = 10$ stations. As before, the 794 CY of sidehill excavation and waste will be moved about one station.

6.1.3 Haul Distances from Mass Diagrams

On large jobs, haul distances are determined by use of mass diagrams which are often roughly plotted on quadrille paper by the estimator. A mass diagram is a curve plotted in connection with a road survey having excavation and embankment quantities. As in the preceding example, cut quantities are considered positive and fill quantities are considered negative.

The ordinates for the mass diagram are calculated from the excavation and the embankment quantities as described in what follows. Since the mass diagram will be used to determine earthwork balances, either the cut or the fill volumes from the road profile must be adjusted by the grading factor prior to being used to calculate the ordinates.

When the grading factor is constant for all of the cut sections, either the cut volumes or the fill volumes may be adjusted. In such cases, however, it is advantageous to adjust the embankment volumes since the mass diagram ordinates will then represent true cut volumes (pay units) and the volumes of cut necessary to make the fills.

When the cut material consists of different types having different grading factors, the cut volumes must be adjusted. In such cases the mass diagram ordinates will represent cut volumes available for fill and the true fill volumes.

In the following examples we assume the grading factor is constant for the entire section considered and adjust the embankment quantities, dividing each of them by the grading factor. Thus, throughout this discussion of mass diagrams, the terms "excavation" or "cut" will always refer to true excavation volumes and the "embankment" or "fill" will always refer to the adjusted quantities.

Typical mass diagrams are shown in Figures 6.2 and 6.3. The abscissas (horizontal distances) on a mass diagram are the roadway stations. The ordinate (vertical distance) at any station is the algebraic sum of the net cut and fill yardages up to that station. Thus, at station 0 the ordinate will be zero, at station 1 the ordinate will be the algebraic sum of the cut and fill volumes between station 0 and station 1, and at station 2 the ordinate will be the algebraic sum of the cut and fill volumes between station 0 and station 2.

If, on a given profile, there is sidehill work or a grade point between, say, stations 8 and 9, a cut volume and a fill volume will both be tabulated on the road profile for that particular interval. Since net volumes (the algebraic sum of the cut and the fill volumes) are used to calculate mass diagram ordinates, the cut volume necessary to balance the adjusted fill volume for that particular interval will not appear on the mass diagram even though the cut must be made, it will appear on the engineer's estimate, and it will be paid for.

On jobs where mass diagrams will be used, therefore, a separate total for such sidehill excavation must be made at the time the engineer's estimate for the roadway excavation is being checked. For any interval, the sidehill excavation will be numerically equal to the smaller of the cut or the adjusted fill volumes for that interval. As a check, the sum of the total sidehill excavation and the total excavation from the mass diagram should equal the total excavation tabulated on the profile.

Figure 6.2. Mass diagram.

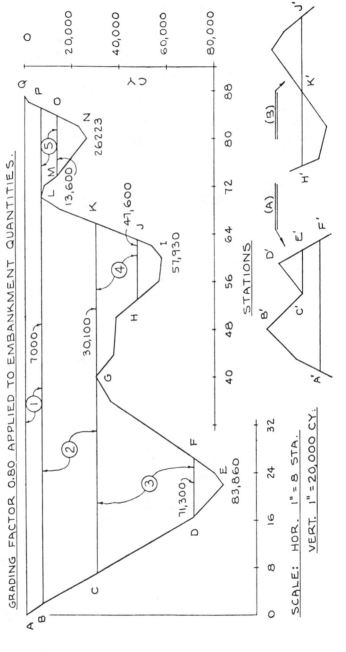

Figure 6.3. Mass diagram (continued).

6.1.3.1 PROPERTIES OF THE MASS DIAGRAM. The essential properties of the mass diagram are:

1. Maximum or minimum points on the curve occur at grade points, where the profile changes from cut to fill or from fill to cut.
2. An upward-sloping curve indicates excavation, and a downward-sloping curve indicates fill. The slopes of the curve vary with the volume of cut or fill per linear foot along the road centerline.
3. A horizontal line intersecting the curve at any two points indicates a balance of cut and fill volumes between the two points of intersection, and the quantities intersepted by any two horizontal lines will balance.
4. Since the abscissas are stations and the ordinates are cubic yards, the areas on the curve represent haul, where haul is defined as volume times distance and is measured in station yards.

6.1.3.2 HAUL DISTANCES. Properties 3 and 4 enable haul distances to be determined. Referring to Figure 6.2, excavation AE will balance fill EB (property 3). Since area AEB is expressed in station yards (property 4), the haul distance between the centers of volume of excavation AE and fill EB is determined by dividing area AEB by the ordinate distance from line AB to the maximum point E. The haul distances for the example given in Table 6.1 will now be determined by the use of a mass diagram. The mass diagram ordinates are calculated in Table 6.2, and the mass diagram is shown in Figure 6.2.

The quantities shown in the embankment and excavation columns of Table 6.2 are assumed to be taken directly from the tabulated quantities on the road profile. The grading factor is assumed to be 0.85, and it is assumed the quantities are tabulated on the profiles in two-station intervals.

TABLE 6.2 Mass Diagram Ordinates

Station	Embankment CY	Embankment 0.85	Excavatio n CY	Net Volume CY	Mass Diagram Ordinate	Sidehill Excavation CY
0					0	—
2	− 200	− 235	+ 400	+ 165	+ 165	235
4	—	—	+ 1000	+ 1000	+ 1165	—
6	—	—	+ 1200	+ 1200	+ 2365	—
8	—	—	+ 1100	+ 1100	+ 3465	—
10	—	—	+ 800	+ 800	+ 4265	—
12	− 300	− 353	+ 300	− 53	+ 4212	300
14	− 800	− 941	—	− 941	+ 3271	—
16	− 1000	− 1177	—	− 1177	+ 2094	—
18	− 800	− 941	—	− 941	+ 1153	—
20	− 1100	− 1294	—	− 1294	− 141	—
22	− 200	− 235	+ 400	+ 165	+ 24	235
Total	− 4400	− 5176	+ 5200	+ 24		770

Balancing line *AB* is positioned on the mass curve so as to be exactly ten stations in length for reasons discussed in Section 6.1.3.3. The excavation from *A* to *E* will exactly balance the embankment from *E* to *B,* and the excavation from *C* to *A* will exactly balance the embankment from *B* to *D.*

The scale of the mass diagram is two stations per inch horizontally and 1000 CY per inch vertically, which equals 2000 station yards per square inch. The area under the curve and above line *AB* is measured with a planimeter or by scaling and is 6.38 square inches, or 12,760 station yards. The ordinate to line *AB* is scaled and equals 2210 CY. The ordinate from line *AB* to the maximum point *E* is 4265 − 2210 = 2055 CY. The haul distance between the centers of volume of excavation *AE* and embankment *EB* is 12,760/2055 = 6 stations.

Similarly, the area between lines *AB* and *CD* measures 15.66 square inches, or 31,320 station yards, and the haul distance between the centers of volume of excavation *CA* and embankment *BD* is 31,320/2210 = 14 stations.

By addition, the area under the curve and above line *CD* = 44,080 station yards, and the haul distance between the centers of volume of excavation *CE* and embankment *ED* = 44,080/4265 = 10 stations.

Referring to the mass diagram, we note that the 141 CY of embankment to the left of station 20 is balanced by the 141 CY of excavation to the right, the haul for this material is about one station, and 24 CY of material will be excavated and wasted.

The sidehill excavation from Table 6.2 is 770 CY, and a summary of the pertinent information for the section of roadway being considered is:

> 2055 CY haul 6 stations
> 2210 CY haul 14 stations
> <u>935</u> CY haul 1 station ± (770 + 141 + 24 = 935)
> 5200 CY total

6.1.3.3 OVERHAUL. On some jobs there is an item for roadway excavation and an item for overhaul. In such cases, the roadway excavation item, as set forth in the specifications, will include the costs of excavating, placing, and hauling the material up to the limit of free haul, which is normally 10 stations. The overhaul item then includes the costs of hauling the material beyond the free-haul limit, and the unit of measure for the item is usually station yards.

Mass diagrams are always used to determine the amount of the overhaul item. In the preceding example, the balancing line *AB* was set at the limit of free haul, which is 10 stations. The average haul distance for excavation *CA* when moved to embankment *BD* was 14 stations, the overhaul distance is 14 − 10 = 4 stations, and the total overhaul for the 22-station section is 2210 CY × 4 stations = 8840 station yards.

Figure 6.3 illustrates a typical mass diagram for a section of freeway. Lines *DF,* *HJ,* and *MO* are 10 stations long, which is the free-haul distance. Thus, excavations *EF, IJ,* and *NO* balance fills *DE, HI,* and *MN,* and none of this material is moved more than 10 stations. The ordinates to the free-haul limit lines are scaled and written on the diagram as shown.

Line AQ is the horizontal axis for the diagram, and lines BP and CK are balancing lines drawn to determine haul distances. Since these lines are drawn through the maximum points G and L, the ordinates to them are taken from the table of ordinates calculated for the diagram. As before, cuts QP, PO, LK, KJ, and GF balance fills AB, LM, BC, GH, and CD.

Table 6.3 shows the calculations required to determine the overhaul distances. The ordinate distances for each of the five sections are listed in column 2, and the areas of the sections are listed in column 3.

Station yards of haul are listed in column 4, and these quantities are obtained by multiplying the areas in column 3 by 160,000 station yards per square inch, which is the scale of the diagram. The stations of haul listed in column 5 are obtained by dividing the station yards of haul by their respective ordinate distances.

The stations of overhaul listed in column 6 are the stations of haul less the free-haul distance, which is 10 stations. Since the grading factor, as noted on the diagram, was applied to the embankment quantities, the ordinates listed in column 2 represent true cut yardages. The station yards of overhaul listed in column 7 are obtained by multiplying these ordinates by their respective stations of overhaul.

As discussed previously, the grading factor must be applied to the cut volumes when the grading factor varies for the different sections of cut. In such a case, the ordinates listed in column 2 would be reduced by the grading factor and would not represent true cut yardages. The areas plotted from such ordinates would also be reduced by the same amount, but the values calculated for stations of haul and for stations of overhaul would not be affected. To determine the station yards of overhaul in such a case, however, the adjusted ordinate is divided by the grading factor to obtain the true cut volume, which is then multiplied by the stations of overhaul.

Two special cases which often occur in connection with mass diagrams are also illustrated in Figure 6.3. Figure 6.3A shows a double loop within the free-haul limit, represented by line $A'F'$. In such a case, the total yardage of free haul is

TABLE 6.3   Overhaul[a]

(1) Section No.	(2) Ordinate CY	(3) Area (sq in.)	(4) SYH	(5) SH	(6) SOH	(7) SYOH
1	7,000	3.709	593,400	84.77	74.77	523,400
2	23,100	9.051	1,448,000	62.68	52.68	1,217,000
3	41,200	5.270	843,200	20.47	10.47	431,400
4	17,500	1.878	300,500	17.17	7.17	125,500
5	6,600	0.492	78,700	11.92	1.92	12,700
					Total	2,310,000

[a]Abbreviations. SYH, station yards of haul; SH, stations of haul; SOH, stations of overhaul; SYOH, station yards of overhaul.

indicated by the sum of the ordinate from line $A'F'$ to B' and the ordinate from line $C'E'$ to D'.

Figure 6.3B represents a pair of loops, and $H'J'$ is a balancing line where both $H'K'$ and $K'J'$ exceed the free-haul distance. In such a case, minimum overhaul between points H' and J' will occur when the balancing line is positioned so that $H'K'$ and $K'J'$ are equal.

6.1.4 Unit Costs for Roadway Excavation

Many earthwork contractors bid roadway excavation strictly on an overall comparison basis. The job under consideration is compared with past, completed work of a similar nature, regarding quantities, type of material, and haul lengths, and the overall unit costs realized on the past work are adjusted for comparison, to reflect the wage and rental rates applicable to the new job.

An overall unit cost for the job being bid is then selected based on the contractor's judgment, the adjusted overall unit costs from the past jobs, and a review of the recent bid prices for similar work as reported in the daily construction services. In most cases, such contractors must wait until the roadway excavation is completed to find out whether money was made or lost on the item.

Although many such contractors manage to stay in business, all successful contractors know of instances where large amounts of money were lost or where bankruptcy occurred because the job bid was not analyzed in sufficient detail to reveal major and costly differences between it and the work with which it was compared.

A far better procedure requires that the roadway excavation item be broken down into similar segments of work, that each of these segments be analyzed, and that a unit cost be derived for each segment based on estimated production rates for the equipment that will be used. These derived unit costs are then compared with the adjusted unit costs realized on past work for similar operations; where appreciable differences are observed, the derived unit costs are reviewed and revised or justified.

The contractor who follows this procedure will, if awarded the job, know exactly what production rates must be achieved for each segment, can intelligently monitor and control the work during construction, and will sleep better at night.

Unlike, say, the structure concrete item, it is not possible to precisely determine the final, on-completion amount of roadway excavation in advance. This results from the inherent imprecision in the selection of the grading factor and in the method used to calculate the volumes from additional compaction, subsidence, and the permitted tolerances in grades and in the location of side slopes.

As discussed in Section 1.6.2.2 and as illustrated in Chapter 8, it is not prudent to mark up the unit cost for an item that may underrun the quantity shown on the engineer's estimate. Therefore, the roadway excavation item is normally bid at or very close to the contractor's bare unit cost for doing the work, and many contractors bid the item at their bare unit cost for excavating, loading, hauling, and compacting the material. In such cases, the fixed costs for moving equipment on and off, for making the grade (which depends on areas which can be precisely deter-

mined), and for definite quantities of included work such as curb or pavement removal will appear, together with the markup for the item, in the price bid for one
or more of the other items. The items so marked up will be lump sum items or
items for which the engineer's estimate can be checked during the takeoff. Such
items should be early items, and the item for clearing and grubbing, which is normally bid lump sum, is frequently used to carry the markup for items whose final
quantities are indeterminate or which will underrun the engineer's estimate.

6.1.5   Wheel Tractor-Scraper Excavation

Different makes and models of wheel tractor-scrapers operate fully loaded under
typical job conditions at about the same speeds (Figure 6.4). The top speed, loaded,
is about 30 miles per hour for both standard and elevating scrapers except that the
smaller sizes of elevating scrapers have a top speed of about 22 miles per hour.
The average travel speed on a given job depends on the condition of the haul road,
its percentage of grade, and the length of haul.

   The overall costs for scraper excavation are greatly affected by the quantities
involved. On small jobs, the fixed costs are high with respect to the volume of
material to be excavated, and the work will normally be completed before a high
production string of equipment could be organized.

   When the quantities are large, the cost estimate should always be based on the
contractor's own cost and production records from past jobs. The various factors
affecting wheel scraper cycle times are discussed in what follows, and empirical
formulas are given for the times required to complete the components of the cycle
times. Although these formulas fit the production data from a number of jobs having

Figure 6.4. Caterpillar wheel tractor-scraper 28-CY struck, 38-CY heaped.

average conditions and average operators working under first-class supervision, the data used by different contractors for calculating scraper cycle times vary. The formulas are presented primarily to illustrate the factors to be considered in developing production records and estimating scraper excavation costs. These formulas should not be blindly used by inexperienced estimators.

The cost to excavate and haul with rubber-tired scrapers depends upon their cycle times:

$$T = L + H + U \qquad (6.1)$$

where

T = cycle time in minutes
L = loading time in minutes
H = hauling time in minutes
U = unloading time in minutes

The loading time will depend on the size of the scraper and the number and size of the push tractors used. Based on a 60-minute hour and assuming average working conditions, the loading times will average about as follows:

$$L_1 = \frac{V}{15} \text{ but not less than 1.0 minute} \qquad (6.2)$$

or

$$L_2 = \frac{V}{25} \text{ but not less than 0.8 minute} \qquad (6.3)$$

where

L_1 = loading time in minutes when one 270-HP push tractor is used
L_2 = loading time in minutes when two 270-HP push tractors or one 385-HP push tractor are used
V = number of bank yards per load

Two 385-HP-plus push tractors are often used with scrapers having heaped capacities greater than about 30 CY, and in such cases formula 6.3 applies. When elevating (self-loading) scrapers are used, the loading time, based on a 60-minute hour, will be about 1.0 minute, and no push tractors are required.

Based on a 60-minute hour, average working conditions, and well-maintained haul roads that are essentially level with no prolonged grades, average travel speeds for wheel tractor-scrapers may be roughly estimated by the following formula:

$$S = 445 + 0.114D_R \qquad (6.4)$$

where

D_R = round-trip haul distance in feet

S = haul speed in feet per minute for values of D_R ranging from 2000 to 14000 feet

When D_R is less than 2000 feet, the scrapers will cycle at an effective rate of 11 cycles per hour, as discussed in Section 6.1.6.

When the loaded scrapers must travel up a grade, the average round-trip travel speeds will be about

$$S_{4G} = 1100 + 0.01D_R \qquad (6.4a)$$

$$S_{6G} = 980 + 0.01D_R \qquad (6.4b)$$

$$S_{8G} = 750 + 0.01D_R \qquad (6.4c)$$

where

S_{4G} = haul speed in feet per minute (4 percent grade) for values of D_R ranging from 7000 to 14,000 feet

S_{6G} = haul speed in feet per minute (6 percent grade) for values of D_R ranging from 5000 to 14,000 feet

S_{8G} = haul speed in feet per minute (8 percent grade) for values of D_R ranging from 3000 to 14,000 feet

When D_R, on the grades, is less than the preceding lower limits, formula 6.4 applies.

In all cases, the average hauling time, in minutes, is

$$H = \frac{D_R}{S} \qquad (6.5)$$

Based on a 60-minute hour, the unloading time will be about

$$U = \frac{V}{90} \text{ but not less than 0.2 minutes} \qquad (6.6)$$

where

U = unloading time in minutes

V = number of bank yards per load

The number of scrapers used on a given haul cannot exceed the production rate for the push tractors used. The number of scrapers that can be loaded by one push tractor or by one pair of push tractors is

$$N = \frac{T}{P} \tag{6.7}$$

where

N = number of scrapers
T = cycle time for scrapers in minutes
P = cycle time for pushers in minutes

The cycle time for the pushers will be about

$$P_1 = \frac{V}{12} \text{ but not less than 1.5 minutes} \tag{6.8}$$

or

$$P_2 = \frac{V}{18} \text{ but not less than 1.1 minutes} \tag{6.9}$$

where

P_1 = pusher cycle time when one 270-HP tractor is used
P_2 = pusher cycle time when two 270-HP tractors or one 385-HP tractor are used
V = number of bank yards per load

Production rates determined by the preceding formulas will be based on a 60-minute hour. As illustrated in the examples which follow, these production rates are adjusted by a 45- or a 50-minute hour when deriving the unit costs to be used on the estimate sheet.

6.1.6 Example: Roadway Excavation—Small Jobs

The roadway excavation quantities shown in Table 6.2 and Figure 6.2 are typical for the roadwork required when a flat slab bridge or a box culvert is constructed. The quantities are summarized again as follows:

	2055	CY haul 6 stations
	+ 935	CY haul 1 station
or	2290	CY haul less than 10 stations
and	2210	CY haul 14 stations

The hourly equipment rental and wage rates are assumed as follows:

Scraper 10/12.5 CY struck/heaped	$48
Bulldozer 270 HP	$60
5 × 5 sheepsfoot roller	$ 7
Blade 125 HP	$39
Water truck 3000 gallons	$30
Grade foreman	$18.50
Grade checker	$16.10

In practice, the applicable rental and wage rates are written at the top of the worksheet used to determine unit costs, and this worksheet is kept immediately behind the estimate sheet for the roadway excavation item.

The cycle time for work within the free-haul limit is slow. The area is congested since the compaction equipment is working close to the cut, and the distances are too short to permit high speeds. On both large and small jobs, the effective production rate for wheel tractor-scrapers working within the free-haul limit will be about 11 cycles per hour.

Assuming the material to be a sandy loam and that one 270-HP pusher will be used, we conclude that the number of bank yards carried per load will be about $(10 + 12.5) \times \frac{1}{2} = 11.2$ BCY, which is equivalent to a load factor of 0.90 applied to the heaped capacity of the scraper. Thus, one scraper will excavate and haul $11 \times 11.2 = 123$ BCY per hour, or 984 BCY per 8-hour day. Considering the quantities involved, two scrapers will be used.

One grade foreman will be required together with two grade checkers: one at the cut and one at the fill. Since the production rate is low and the fill is close to the cut, the haul roads will be maintained by the blade and the water truck used on the fill at no additional cost.

The unit costs for excavating the material within the free-haul limit are determined as follows:

Hourly rental cost:

Push tractor 270 HP	$ 60
2 scrapers at $48	96
Blade 125 HP	39
Tractor plus sheepsfoot	67
Water truck	30
Total	$292

Unit rental cost:

$$\frac{\$292}{2 \times 123} = \$1.19 \text{ per CY}$$

Unit labor cost:

$$\frac{18.50 + (16.10 \times 2)}{2 \times 123} = \$0.21 \text{ per CY}$$

To estimate the unit costs for the work outside the free-haul limit, the cycle time must be determined. From Equation 6.2 we have

$$L_1 = \frac{11.2}{15} = 0.75, \text{minimum} = 1.0 \text{ minutes}$$

From Equations 6.4 and 6.5, the travel time for a round-trip haul of 28 stations is

$$S = 445 \times 0.114 \times 2800 = 764 \text{ feet per minute}$$

and

$$H = \frac{2800}{764} = 3.7 \text{ minutes}$$

This haul time assumes that the haul road is essentially level. From Equation 6.6, the unloading time is

$$U = \frac{11.2}{90} = 0.12, \text{minimum} = 0.2 \text{ minutes}$$

and the total cycle time is $1.0 + 3.7 + 0.2 = 4.9$ minutes.

The scrapers carry 11.2 bank yards per load, and the effective production rate based on a 50-minute hour is

$$\frac{50}{4.9} \times 11.2 = 114 \text{ CY per hour}$$

Two scrapers will be used, and the unit costs for equipment and for labor are:

$$\text{Equipment} \qquad \frac{292}{2 \times 114} = \$1.28 \text{ per CY}$$

$$\text{Labor} \qquad \frac{18.50 + (2 \times 16.10)}{2 \times 114} = \$0.22 \text{ per CY}$$

As before, the haul roads will be maintained by the blade and the water truck used on the fill at no additional cost. In both instances we assume that the water supply is on the job, and the costs to develop the water supply are covered under a separate job item. In this example only two scrapers are used, the production rate is low, and it is not necessary to check the cycle time for the push tractor nor the production rates for the compaction and watering equipment.

6.1.6.1 UNIT COST FOR OVERHAUL. As illustrated in the preceding example, the costs to excavate and move dirt beyond the free-haul distance are based on the actual hauls involved, and a unit cost for overhaul is not considered. In the example, the scrapers are traveling at an average speed of 764 feet per minute = 382 stations per hour based on a 50-minute hour. The unit cost for overhaul would be

$$C = \frac{2R}{VS}$$

where

C = unit rental cost for overhaul in dollars per station yard
R = hourly rental rate for the hauling unit
V = bank yards carried per load
S = average haul speed in stations per hour

The factor 2 is required in the numerator since the overhaul item is based on one-way haul distances, while the hauling units must make round trips. In our example, the unit rental cost for overhaul based on a 50-minute hour would be

$$\frac{48 \times 2}{11.2 \times 382} = \$0.0224 \text{ per station yard}$$

This calculation does not reflect the cost for maintaining haul roads in the overhaul area, and the total cost for such maintenance is included in the estimated cost for the roadway excavation.

Since the estimated costs for roadway excavation include the costs for overhaul, no cost for overhaul is shown in the cost column of the estimate summary sheet. The unit price bid for overhaul is selected somewhat arbitrarily. If the pay quantity for the overhaul item is based on field calculations by the resident engineer, the pay quantity may underrun the engineer's estimate, and the item would be bid at a unit price below the theoretical unit cost. If the pay quantity for the item is specified to be the quantity shown on the engineer's estimate, the item is, in effect, a lump sum item and would be bid at a unit price above the theoretical unit cost. The estimated cost for the overhaul item is deducted from the estimated cost for the roadway excavation, which includes overhaul, to determine the bare unit cost of the roadway excavation item alone.

As illustrated in the sample estimates in Chapter 8, the difference between the balanced unit price and the unit price actually bid for any item is automatically absorbed by the unit prices bid for some of the other items since the totals for the cost columns and the price bid columns of the estimate summary sheet must balance.

6.1.7 Example: Roadway Excavation—Large Jobs

On jobs having large quantities of roadway excavation, higher production rates and greater efficiencies can be achieved. The procedures for estimating the costs of such work are similar to those illustrated in the preceding example except that care must be taken to balance the pushing, hauling, and compaction operations; haul roads will normally require additional equipment for maintenance purposes; and the number of water trucks or tankers must be determined.

Consider the portion of freeway relating to Figure 6.3 and Table 6.3, which is 87 stations in length. In addition to the 95,400 CY to be overhauled, as shown in column 2 of the table, there are $12,560 + 10,330 + 12,623 = 35,513$ CY to be excavated and placed within 5 limits of free haul at points E, I, and N. Although the portion of freeway shown involves fairly heavy cut, it is only 1.6 miles long. Considering that typical freeway jobs average from 5 to 12 miles in length, it is clear that maximum efficiency must be attained on such work.

Using the quantities in Table 6.3, we have:

Section	CY of Cut	Stations of Haul
1	7,000	84.8
2	23,100	62.7
3	41,200	20.5
4	17,500	17.2
5	6,600	11.9

Combining sections 3 and 4, whose haul lengths are about the same, we use the weighted average haul for the two sections and have:

Section	CY of Cut	Stations of Haul
1	7,000	84.8
2	23,100	62.7
3 + 4	58,700	19.5
5	6,600	11.9

The hourly rental and wage rates are assumed to be:

Scraper, 21 CY struck, 30 CY heaped	$77
Bulldozer 270 HP	$60
Bulldozer 385 HP	$79
5 × 5 sheepsfoot roller	$7
Blade 125 HP	$39
Blade 150 HP	$40
Water tanker 5000 gallons	$35
Grade foreman	$18.50
Grade checker	$16.10
Laborer	$13.48

Assuming the material to be a sandy loam and that a 385-HP tractor will be used as a pusher, we conclude that the scrapers will carry $1.26 \times 21 = 26.4$ bank yards per load, which is equivalent to a load factor of 0.88 applied to the heaped capacity. From Equation 6.3 we have

$$L_2 = \frac{26.4}{25} = 1.1 \text{ minutes}$$

and from Equation 6.6

$$U = \frac{26.4}{90} = 0.3 \text{ minutes}$$

The preceding times will be constant for all of the excavation and are based on a 60-minute hour.

Since sections $3 + 4$ contain the largest portion of the excavation, the scraper cycle time is estimated for this section first. Assuming essentially level haul roads, we have, from Equation 6.4,

$$S = 445 + 0.114 \,(19.5 \times 100 \times 2) = 890 \text{ feet per minute}$$

and from Equation 6.5 we find

$$H = \frac{19.5 \times 100 \times 2}{890} = 4.4 \text{ minutes}$$

The cycle time for sections $3 + 4$ based on a 60-minute hour is

$$1.1 + 4.4 + 0.3 = 5.8 \text{ minutes}$$

Based on a 50-minute hour, the production rate per scraper will be

$$\frac{50}{5.8} \times 26.4 = 227 \text{ BCY per hour}$$

From Equation 6.9 the cycle time for the pusher is found to be

$$P_2 = \frac{26.4}{18} = 1.5 \text{ minutes}$$

and the maximum number of scrapers that can be loaded with the one pusher is, from equation 6.7,

$$N = \frac{5.8}{1.5} = 3.9$$

If three scrapers are used, the production rate for sections 3 + 4 based on a 50-minute hour will be

$$3 \times 227 = 681 \text{ CY per hour}$$

or

$$681 \times 8 = 5450 \text{ CY per 8-hour day}$$

The time required to complete sections 3 + 4 will be

$$\frac{58,700}{5450} = 11 \text{ days}$$

By increasing the cycle time for the scrapers to 6.0 minutes, four scrapers may be used, but there will be some waiting, in the cut, for the pusher. Based on a 50-minute hour, the production rate for four scrapers will be

$$\frac{50}{6.0} \times 26.4 \times 4 = 880 \text{ CY per hour}$$

or

$$880 \times 8 = 7040 \text{ CY per 8-hour day}$$

and 9 days will be required for completion.

Using the same procedures, the cycle times and the production rates are determined for sections 2, 1, and 5, and the data are entered in Table 6.4.

In addition to sections 1 through 5, there are 35,513 CY of material within the free-haul limit, as shown on the mass diagram (Figure 6.3). This quantity should be increased by the amount of the sidehill excavation taken from the road profile, and for this example we assume the total material within the free-haul limit, including such sidehill excavation, is 36,500 CY.

As previously discussed, the effective production rate for rubber-tired scrapers working within the free-haul limit is 11 cycles per hour per scraper. Thus, the effective (50-minute hour) cycle time is

$$\tfrac{50}{11} = 4.54 \text{ minutes}$$

and the equivalent cycle time based on a 60-minute hour would be

$$4.54 \times \frac{5}{6} = 3.78 \text{ minutes}$$

As previously determined, the pusher cycle time based on a 60-minute hour is 1.5 minutes, and from Equation 6.7 we find

$$N = \frac{3.8}{1.5} = 2.5$$

Two scrapers will be used, and the effective production rate based on a 50-minute hour will be

$$2 \times 26.4 \times 11 = 581 \text{ CY per hour}$$

or

$$581 \times 8 = 4650 \text{ CY per 8-hour day}$$

and 8 days will be required to complete this work. This information is entered in Table 6.4, where we now have an array of the possible production rates, by sections, for various combinations of equipment. Such an array will simplify the final selection of the number of scrapers to use for each section since it may not be possible or feasible to bring in an additional scraper for a short period of time. Before making this decision, the production rates for the compacting and watering equipment should be considered.

The production rates for a double 5 × 5 sheepsfoot roller and for a twin vibratory sheepsfoot roller are entered in Table 6.4.

TABLE 6.4 Possible Production Rates

Section	BCY	Number of Scrapers	Daily Production BCY[a]	Days to Complete
1	7,000	3	3,270	2.1
		4[b]	4,350	1.6
		5	5,440	1.3
2	23,100	3	3,910	5.9
		4[b]	5,220	4.4
		5	6,520	3.5
3 + 4	58,700	3[b]	5,450	10.8
		4	7,040	8.3
5	6,600	2[b]	4,490	1.5
		3	6,740	1.0
Free haul	36,500	2[b]	4,650	7.8

[a]Based on a 50-minute hour and an 8-hour day.
[b]Number of scrapers selected.

Note: Estimated production rates (BCY per day—see text):

Double 5 × 5 sheepsfoot roller 5,600
Twin vibratory sheepsfoot roller 6,400
5000-gallon water tanker 3,200

6.1.7.1 WATER REQUIRED FOR COMPACTION. The amount of water required for compaction is usually estimated from past job records. In the absence of such records an approximation may be made. Assume the following information is known (CF = cubic feet):

Maximum density of the soil	113 pounds per CF
Optimum moisture content	10%
Density at 90% relative compaction	101.7 pounds per CF
Moisture in bank material	2%
Grading factor	0.80

The moisture percentages are determined on a weight basis. The theoretical amount of added water required for compaction is $10 - 2 = 8.0$ percent. Adding 1 percent for field compaction and 5 percent for waste, evaporation, and the water used to maintain haul roads, we estimate that 14 percent of additional moisture will be required. This amounts to

$$0.14 \times 101.7 = 14.2 \text{ pounds per compacted cubic foot}$$

or

$$\frac{14.2 \times 27}{8.33} = 46.0 \text{ gallons per compacted cubic yard}$$

which is

$$46.0 \times 0.80 = 37 \text{ gallons per bank yard}$$

The cost to apply water is based on the following assumptions: The water supply is located near the middle of the job and the cost to develop and to furnish the water supply is included under a separate job item. Semi-trailer sprinkling tankers having a capacity of 5000 gallons will be used (see Figure 6.5 for one with a capacity of 10,000 gallons). The water will be loaded from a portable water tower (Figure 6.6) having a 12,000-gallon tank and an effective discharge rate of 2500 gallons per minute. Finally, we assume water tankers will discharge at a maximum effective rate of 1000 gallons per minute.

Assuming a round-trip haul distance of 1.6 miles and an average tanker speed of 12 miles per hour, the cycle time for the 5000-gallon tankers is estimated as follows:

$$\text{Loading time } \frac{5000}{2500} + 1 = 3.0 \text{ minutes}$$

$$\text{Hauling time } \frac{1.6 \times 60}{12} = 8.0 \text{ minutes}$$

$$\text{Discharge time } \frac{5000}{1000} = \underline{5.0 \text{ minutes}}$$

$$\text{Minimum cycle time:} \quad 16.0 \text{ minutes}$$

Figure 6.5. *Semi-trailer-type sprinkler tank, 10,000 gallons (Southwest Welding and Manufacturing Co., Alhambra, California).*

Figure 6.6. *Hydraulically erected mobile water tower (Southwest Welding and Manufacturing Co., Alhambra, California).*

We estimate the effective cycle time will be 20 minutes per load.

Each tanker will therefore deliver 15,000 gallons per hour, which is sufficient to compact

$$\frac{15000 \times 8}{37} = 3200 \text{ bank yards per 8-hour day}$$

The production rates for the water tankers are noted at the bottom of Table 6.4

Reviewing the information in Table 6.4, the following equipment is selected:

Section 1	4 scrapers
Section 2	4 scrapers
Sections 3 + 4	3 scrapers
Section 5	2 scrapers
Free haul	2 scrapers

Two water tankers will be ample for all sections, and the material will be compacted with a large blade and a double 5 × 5 sheepsfoot roller.

6.1.7.2 MOTOR GRADERS REQUIRED ON THE FILL. The number of motor graders required on the fill is usually based on the estimator's judgment and past experience. A rational approach, however, may be made as follows:

From Table 6.4, the maximum production rate is

$$\frac{5450}{8} = 681 \text{ BCY per 50-minute hour}$$

Assuming a load factor of 0.85, this is equivalent to

$$\frac{681}{0.85} = 801 \text{ LCY per 50-minute hour}$$

Assuming the material will be spread about 4 inches deep by 12 feet wide, we have

$$\frac{0.33 \times 12 \times L}{27} = 801 \text{ LCY}$$

and

$$L = 5500 \text{ feet or 55 stations}$$

where

L = the length of the material spread per 50-minute hour

From Equation 4.10, assuming an average speed of 4.5 miles per hour and that three passes will be required to mix the water with the material, we have

$$P_s = \frac{4.5 \times 52.8}{3} = 79 \text{ stations per 60-minute hour}$$

$$= 79 \times \frac{50}{60} = 66 \text{ stations per 50-minute hour}$$

and the number of motor graders required would be

$$\tfrac{55}{66} = 0.8$$

and one scraper will be used.

6.1.7.3 UNIT COSTS. The daily cost for the loading equipment, a 385-HP bulldozer, is $8 \times \$79 = \$632$. The daily costs for the hauling units are:

2-scraper string $2 \times 8 \times \$77 = \$1232$
3-scraper string $3 \times 8 \times \$77 = \$1848$
4-scraper string $4 \times 8 \times \$77 = \$2464$

The costs for the compacting equipment are:

Blade 150 HP	$40 per hour
Bulldozer 270 HP	$60 per hour
Sheepsfoot roller	$ 7 per hour
Two 5000-gallon tankers	$70 per hour
	$177 per hour

or

$$8 \times \$177 = \$1416 \text{ per 8-hour day}$$

In all cases, the labor will consist of one grade foreman, two grade checkers, one in the cut and one on the fill, plus one laborer, who will be used to keep a load count and a cycle time record for the scrapers. The daily cost for this crew will be

$$8(18.50 + 2 \times 16.10 + 13.48) = \$513.44$$

The unit costs for labor and for equipment are:

Section 1:

Labor $\qquad\qquad \dfrac{513.44}{4350} = \$0.118 \text{ per BCY}$

Rental $\dfrac{632 + 2464 + 1416}{4350}$ = \$1.04 per BCY

Section 2:

Labor $\dfrac{513.44}{5220}$ = \$0.098 per BCY

Rental $\dfrac{632 + 2464 + 1416}{5220}$ = \$0.864 per BCY

Sections 3 + 4:

Labor $\dfrac{513.44}{5450}$ = \$0.094 per BCY

Rental $\dfrac{632 + 1848 + 1416}{5450}$ = \$0.715 per BCY*

Section 5:

Labor $\dfrac{513.44}{4490}$ = \$0.114 per BCY

Rental $\dfrac{632 + 1232 + 1416}{4490}$ = \$0.731 per BCY

Free haul:

Labor $\dfrac{513.44}{4650}$ = \$0.110 per BCY

Rental $\dfrac{632 + 1232 + 1416}{4650}$ = \$0.705 per BCY

6.1.7.4 MAINTAINING THE HAUL ROAD. The cost to maintain the dirt haul roads is estimated as a separate lump sum item on the estimate sheet for roadway excavation. Typically, the equipment allowed for this will be based on past job experience.

Referring to Table 6.4, we note that the excess watering capacity available from the fill tankers working on section 1 is

$$(6400 - 4350) \times \frac{37}{8} = 9500 \text{ gallons per hour}$$

The excess watering capacity available is determined for the remaining sections, and the following array is set up:

Section	Excess Capacity (gallons per hour)	Length of Haul Road (stations)
1	9500	85
2	5500	63
3 + 4	4400	20
5	8800	12
Free haul	8100	10

The preceding haul road lengths are taken directly from column 5 of Table 6.3 except that the weighted average is used for sections 3 and 4.

Assuming, as before, that each haul road water truck will make three trips per hour, it will spend at least $3 \times 5 = 15$ minutes per hour watering at the road. The excess capacity noted for section 1 amounts to two loads per hour, or an additional 10 minutes of watering time per hour. Assuming an average speed while watering of 10 miles per hour, the watering equipment will cover

$$\tfrac{25}{60} \times 10 \times 52.8 = 220 \text{ stations per hour}$$

which is more than two passes per hour over the haul road for section 1. Considering the remaining sections in a similar manner, we find that one haul road water truck plus the available excess capacity from the fill will result in more than two passes per hour on section 2, the excess capacity from the fill will result in 1.9 passes per hour on sections 3 and 4, and the excess capacity from the fill is ample for the remaining two sections.

If one 3000-gallon water truck is used on the haul roads, the total water available for section 1 is

$$3 \times 3 + 9.5 = 18,500 \text{ gallons per hour}$$

or

$$\frac{18,500}{7.48} = 2470 \text{ CF per hour}$$

Assuming the haul road is 36 feet in width, we have

$$85 \times 100 \times 36 \times t = 2470, \qquad t = 0.008 \text{ feet per hour}$$

where

t = depth of water applied per hour

The weight of the water applied per hour is

$$0.008 \times 62.4 = 0.5 \text{ pounds per hour per SF}$$

Assuming the soil weighs 100 pounds per CF, a 6-inch layer will weigh 50 pounds, and we will be applying 1 percent, by weight, of moisture per hour, which will be adequate. Checking the remaining sections in a similar manner, we decide to use one 3000-gallon water truck on the haul roads for sections 1 and 2. Sections 3 and 4, section 5, and the free-haul section will be maintained with the excess capacity from the fill tankers. Referring to Table 6.4, the water truck will be required for 7 days.

One motor grader traveling 4.5 miles per hour with an effective blade width of 10 feet will make one pass over

$$4.5 \times 5280 \times \tfrac{10}{9} \times \tfrac{5}{6} = 22,000 \text{ SY per 50-minute hour}$$

and we will need a minimum of one pass per hour to maintain the haul road. Considering the haul road areas, we decide to use two 125-HP blades on sections 1 and 2 and one blade on sections 3 and 4. The remaining two sections will be maintained by the blade working on the fill. The days of blade time required, from Table 6.4, are

$$(2 + 5)\, 2 + 11 = 25 \text{ days}$$

Assuming hourly rental rates of $30 for the water truck and $39 for the motor grader, the equipment cost to maintain the haul roads is

$$7 \times 8 \times 30 + 25 \times 8 \times 39 = \$9500$$

6.1.7.5 ON-AND-OFF CHARGES FOR EQUIPMENT. The determination of the cost to move the equipment on and off the job is the final item on the estimate sheet for roadway excavation. In addition, an estimated allowance is made for the survey stakes required.

6.1.8 Compacting the Original Ground

Specifications for roadway excavation normally require a relative compaction of 95 percent to be obtained for a minimum depth of 2 to 3 feet below the finish roadway grade for at least the width of the traveled way whether in excavation or in embankment. Thus, in all cut areas and in areas of low embankments, undercutting (additional roadway excavation) is required.

The undercutting is made to a plane 8 inches above the lower limit for the compacted zone, and the unit costs derived for roadway excavation within the free-haul limit will apply to this work. The compaction of the bottom 8 inches is designated as compacting the original ground, and the cost for this work is subject to considerable variation from job to job.

The unit cost for the equipment required to compact original ground is indexed to the total hourly rental rate for the following equipment:

1 bulldozer
1 double 5 $\times$ 5 sheepsfoot roller
1 motor grader
1 water truck

The production rate achieved by this equipment will vary with the sizes of the individual areas involved, the type of material, and the degree of difficulty encountered in mixing the water with the material when it is dry or in aerating and drying it out when it is too wet.

Areas of 30 MSY (thousand square yards) in sandy loam have been compacted at an effective rate of 0.70 MSY per hour, areas of 20 MSY in pure sand have been compacted at an effective rate of 1.2 MSY per hour, and areas of 2 MSY in sandy loam have been compacted at an effective rate of 0.47 MSY per hour. In most cases, no additional labor will be required since the work will be done concurrently with the other roadway excavation.

6.1.9 Imported Borrow

Imported borrow is normally bid as a separate job item. In every case, the cost of the borrow pit site, or the royalty paid for the borrow material, should be the first item on the estimate sheet for imported borrow.

The unit price paid for imported borrow includes the costs of furnishing, loading, hauling, placing, and compacting the material as well as the cost to make the grade on the completed embankment. The unit of measure for payment is usually bank yards excavated or tons. Some agencies will, however, pay for the item on a compacted embankment yard basis.

When off-highway hauling is possible and the one-way haul distance is about 1.25 miles or less, scrapers are normally used. When the import must be hauled over public roads or the one-way haul distance is about 1.50 miles or more, the material is hauled in trucks.

Significant savings are possible when borrow pits can be located close to their related embankments, and jobs are often won by obtaining optimum pit sites. Where land values are high, deeper pits are used. Typical pits range from 10 to 20 acres in area and from 25 to 40 feet in depth.

Prior to making a final decision on a pit site, test holes are drilled to see if the proposed material meets the specifications. When occasional thin layers of unsuitable material are found, a satisfactory product can sometimes be obtained by opening up the pit and loading the remaining material from top to bottom. The material is thus loaded in a downhill direction, which is advantageous, and by cutting across the layers, a mixed product is obtained at no additional cost.

The cycle time for hauling units working out of borrow pits is

$$T = L + P + S + H + U \qquad (6.10)$$

where

T = cycle time in minutes

L = loading time in minutes

P = time required to get in and out of the pit in minutes

S = time spent on scales in minutes

H = haul time in minutes

U = unloading time in minutes

Scales, when used, are located at the pit exit, and the weighmaster is normally furnished by the owner at no cost to the contractor. The time spent on the scales will be about 0.2 minutes per load.

Scales may be rented by the month, and the labor cost to set up and remove a set of scales will be about $1500 based on $K_{6+2} = 1000$, and the equipment cost will be about $750 based on $K_{Cr} = 75$. In addition, from 5 to 7 CY of concrete will be required, and the truck time for hauling the scales on and off the job will be additional.

Careful consideration must be given to the time required to get in and out of the pit. This time will depend on the pit size, its depth, the type of the pit material, the pit ramp grade, and the type of hauling units used.

Shallow pits pose no problem for scrapers, but the haul road in the pit must be maintained and its length included when estimating the hauling time for the scrapers. Shallow pits may require lower truck speeds when the pit material is sandy or otherwise unsuitable for truck travel. In addition, when trucks are used, the pit haul roads must be maintained even though the out-of-pit haul is over public roads.

The average haul speeds for equipment working in deep pits with ramps will be about 400 feet per minute for scrapers and about 10 to 12 miles per hour for trucks.

Except for the additional consideration of the possible loss of time in the pit, the procedures for estimating the costs for borrow pit excavation with scrapers are the same as those illustrated previously for roadway excavation. When trucks are used, the unloading time requires careful analysis and will include the time spent hauling over the embankment in progress.

Although haul roads over original ground or completed embankments can be well maintained, the top layer of the fill in progress will be composed of loose or partially compacted material which will result in lower truck speeds. To cut the unloading time, sets of doubles are normally used to haul import when large quantities are involved. These consist of a truck-tractor pulling two bottom-dump trailers of equal size, the front trailer being a semi-trailer. Such units have a shorter turning radius than a semi of the same capacity, can hit the fill at high speeds, and can dump on the fly. When import is hauled in end-dump trucks, an additional 3 to 5 minutes per load is required for the dumping time.

The estimation of the time spent on the fill requires job experience, and the time will depend on the length and grade of the fill, the type of material being placed, and the possible approach speeds.

Wheel loaders are frequently used to load trucks with imported borrow. When high production rates are necessary, large loaders are required, and these are seldom available for rental.

6.1.10 Haul Distances for Imported Borrow

The determination of haul distances for imported borrow is usually simple since all the material from a given pit will normally go into one or two large embankments. The centers of volume for the embankments are often determined by inspection since the embankment quantities are tabulated at uniform intervals along the road profile. When necessary, centers of volume for embankments are determined by the methods illustrated in Section 6.1.2.

When a single borrow pit is located at the side of a road job requiring import along its entire length, haul distances must be determined by dividing the job into two sections, the point of division being the point where the haul road from the pit enters the job right-of-way. The haul lengths will then be the distances, measured along the haul roads, from the center of volume of the pit to the centers of volume of those embankments located along each of the two job sections. In such cases, it is incorrect to assume that the average haul length for all the import is the distance from the center of volume of the pit to the center of volume of all of the embankments on the entire job.

6.1.11 Belt Loaders

Specially designed conveyor belts fed by two to four bulldozers are widely used when large amounts of earth must be loaded into trucks (Figures 6.7 and 6.8). The belt loader is a semi-portable, self-contained unit from 50 to 60 feet long with a belt width of 4, 5, or 6 feet. The loader is equipped with a low-profile dozer trap and a heavy-duty plate feeder. When needed, a vibrating screen or grizzly may be

Figure 6.7. Kolman belt loader, 5-foot belt, 50 feet long (Kolman Division, Athey Products Corp., Sioux Falls, South Dakota).

Figure 6.8. *Kolman belt loader. Screening oversize material and loading bottom-dump semi-trailers (Kolman Division, Athey Products Corp., Sioux Falls, South Dakota).*

added to scalp off oversize material, and this important capability is unique to the belt loading method.

Provision must be made, in all cases, to keep the truck loading area free of spillage or scalped material. A small rubber-tired tractor equipped with a back blade is used to remove spillage since it is cheaper and more maneuverable than a motor grader.

The output capacity of the belt loader depends on the number of bulldozers used to feed it, the capacity of the belt, the size of the haul units, and the ability of the belt operator, who must stop the belt after each haul unit is loaded. Assuming the belt loader itself has ample capacity, the output of the system may be increased by as much as 33 percent for short periods of time. This is accomplished by decreasing the haul distance for the bulldozers and is a second important and unique characteristic of the belt method of loading. A belt loader can get a string of waiting trucks loaded and back on their normal cycles in less time than is required by any other method of loading.

Belt loaders are moved back into the cut as the work progresses. This is done after the normal work shift; two bulldozers and two workmen will make the move in about 2 hours.

6.1.11.1 BELT LOADER EXAMPLE. The following conditions are assumed:

1. The pit is an average of 600 feet × 1000 feet in area by 25 feet deep and contains 550,000 bank yards. The material is sandy loam weighing 1.50 tons per bank yard.
2. The haul road is 3.2 miles long, one way, from pit exit to the beginning of the fill, and no public roads are used.
3. Average haul distance on the fill is 900 feet.

4. The material is hauled in sets of doubles having a capacity of 24 tons per set.
5. Scales are not required.

Based on past job records, a belt loader with a 5-foot-wide belt will load 24 tons of material into a set of doubles in 1 minute, which is equivalent to 1440 tons of material per 60-minute hour. Thus, referring to Equation 6.10, we have

$$L = 1.0 \text{ minutes}$$

One 270-HP bulldozer with a straight blade should move about 340 LCY per 60-minute hour on a 150-foot haul. Assuming a load factor of 0.80, this is equivalent to

$$340 \times 0.80 = 270 \text{ bank yards per hour}$$

or

$$270 \times 1.5 = 400 \text{ tons per hour}$$

Four 270-HP bulldozers will be used. The estimated production rate for the four bulldozers is 1600 tons per 60-minute hour.

Assuming the haul road is well maintained and has no excessive grades or sharp turns, the average haul speed for the doubles will be about 27 miles per hour. Referring to equation 6.10, we have

$$H = \frac{2 \times 3.2 \times 60}{27} = 14.2 \text{ minutes based on a 60-minute hour}$$

Assuming the trucks average 10 miles per hour on the fill and they dump on the fly, we have

$$U = \frac{2 \times 900 \times 60}{5280 \times 10} = 2.1 \text{ minutes based on a 60-minute hour}$$

Assuming the in-pit haul to average about 600 feet one way and an average truck speed of 12 miles per hour, we have

$$P = \frac{2 \times 600 \times 60}{5280 \times 12} = 1.1 \text{ minutes based on a 60-minute hour}$$

From Equation 6.10 the total cycle time based on a 60-minute hour is

$$T = 1.0 + 14.2 + 2.1 + 1.1 = 18.4 \text{ minutes}$$

and the production rate per set of doubles will be

$$\frac{60}{18.4} \times 24 = 78.3 \text{ tons per 60-minute hour}$$

The number of trucks required is

$$\frac{1440}{78.3} = 18.4 \text{ trucks}$$

and 19 trucks will be used. Both the production rates used to determine the number of trucks required are based on 60-minute hours.

The effective production rate based on a 50-minute hour will be

$$1440 \times \tfrac{50}{60} = 1200 \text{ tons per hour}$$

which, in this case, is equivalent to

$$\frac{1200}{1.50} = 800 \text{ bank yards per hour}$$

Based on the pit size and the production rate, we conclude that the belt will be moved back into the pit face on 5-day intervals, which is equivalent to one move for each 48,000 tons.

The equipment required to water and compact the fill and the equipment required to maintain the haul road are determined by the procedures outlined previously in the example for roadway excavation.

The labor required for the entire operation will be one grade foreman, one grade checker in the pit, one grade checker on the fill, and one laborer at the belt loader who will keep a log of the truck cycle times.

The unit costs for imported borrow are determined as shown in the preceding example for roadway excavation. Since only one haul road is involved, the equipment cost to maintain the haul road may be based on a unit cost per ton of material hauled.

Thus, if the rental cost for the haul road equipment were $144 per hour, the unit cost for haul road maintenance would be

$$\frac{144}{1200} = \$0.12 \text{ per ton}$$

based on the effective production rate of 1200 tons per hour. Since the total length of the haul road, including the section in the pit, is 3.3 miles, we have an hourly haul of

$$1200 \times 3.3 = 3960 \text{ ton-miles}$$

and for comparison when estimating future work, the haul road maintenance cost might also be expressed as

$$\frac{144}{3960} = \$0.0364 \text{ per ton-mile}$$

6.1.12 Making the Grade

The cost to make the final grade will depend on the degree of care exercised in making the rough grade for the area being considered and on the allowable tolerances for the final grade. The rough grade for the roadbed in cut or embankment should be made to about ±0.3 foot at the time the work is done, and care should be taken to balance the plus and minus areas, with any error on the plus side.

The base unit cost for the equipment required to make grade is, for estimating purposes, indexed to the hourly rental costs for the following equipment, and these rental costs are assumed as shown:

1 blade	$ 39
1 water truck	30
1 self-propelled pneumatic roller	36
$\frac{1}{2}$ of 9-CY elevating scraper $46 × $\frac{1}{2}$	23
	$128

The sum of the preceding hourly rental costs is designated as K_{MG-R}.

The base unit cost for the labor required to make grade is, for estimating purposes, indexed to the hourly wage rates for the following crew, and these wage rates are assumed as shown:

1 grade foreman	$18.50
1 grade checker	16.10
1 guinea hopper	13.70
1 laborer	13.48
	$61.78

The sum of the preceding hourly wage rates is designated as K_{MG-L}. On jobs where the quantity of grade to be made amounts to about 16 MSY or more, the equipment and crew will make grade on the roadbed at an effective rate of about 1 MSY per hour assuming tolerances of ±0.1 foot on the shoulders and ±0.05 foot on the subgrade.

Thus, based on $K_{MG-R} = 128$, the base unit equipment cost for making grade on the roadbed is $128 per MSY, and based on $K_{MG-L} = 61.78$, the base unit labor cost is $61.78 per MSY. These unit costs assume all areas are large enough to permit free and unrestricted use of the equipment.

Lower production rates are realized when smaller areas are involved. Thus grade will be made on areas of 3 MSY at a rate of about 0.45 MSY per hour, and on areas of 10 MSY the rate will be about 0.65 MSY per hour. When areas as small as 0.3 MSY are involved, the production rate will be about 0.35 MSY per hour.

When areas are confined and provide a minimum of space for the equipment, the unit costs for making grade increase, and the estimated amount of the increase will depend on the estimator's judgment.

The grade tolerance for medians is normally ±0.2 foot for the general area and ±0.1 foot for drainage inverts. The crew and equipment listed in the preceding will make grade on median areas of 16 MSY or more at an effective rate of about 1.5 MSY per hour.

6.2 BASE COURSES AND PAVEMENTS

Aggregate subbase, aggregate base, and cement-treated base are commonly used as base courses in the structural sections of roadways. The thicknesses and types of base courses required are shown on typical roadway cross sections such as the one illustrated in Figure 6.1.

The unit price paid for the base course includes the cost to furnish, spread, and compact the material and the cost to make the grade on top of it. The payment for base courses is based on the plan dimensions for the structural section. Thus, when material is paid for on a cubic yard basis, the unit of measure is compacted cubic yards.

Portland cement concrete pavement may be placed by general contractors but is often installed on a subcontract basis. Asphalt paving is normally installed by subcontractors who have permanent mixing plants and aggregate sources in the area.

6.2.1 Aggregate Subbase (AS)

Aggregate subbase is a sandy material having a specified gradation and, when required, is the bottom course in the structural section. Aggregate subbase is obtained from borrow pits, commercial rock plants, or in some instances, the roadway excavation. When obtained from borrow pits, the costs to furnish the material to the grade are estimated by the methods outlined earlier in this chapter except that the dumping area will be relatively restricted and lower production rates will be realized.

Typical sections of aggregate subbase will range from 0.5 to 1.5 feet in thickness. The material is dumped on the grade, spread with a motor grader, and compacted to a relative compaction of 95 percent. When the section is more than 0.5 feet in thickness, the specifications may require the material to be placed in two or more layers of approximately equal thickness, and the compacted thickness of an individual layer, in such cases, cannot exceed 0.5 feet. Typically, a small wheel tractor pulling a set of discs will be required to mix the water with the subbase during compaction.

When the material can be hauled with scrapers, a top production rate of 300 compacted cubic yards per hour may be maintained on large jobs where the structural section is about 50 feet wide or more. The costs for hauling with scrapers were discussed in preceding sections.

When the material is hauled by trucks, the maximum production rate will be about 300 tons per hour, and this rate may not be practicable where the hauls are long because of the number of trucks required. When aggregate subbase, in relatively small amounts, is furnished by commercial plants, the loading may be done with plant equipment, and a total of about 10 minutes should be allowed for loading and unloading the trucks.

When large amounts of material are required from commercial plants, the contractor should do his own loading or get a commitment from the plant concerning the loading rate if the plant loaders are used. In such cases the truck cycle times are estimated as shown in Section 6.2.2. In general, whenever trucks are used, a load counter should keep a log of the truck cycle times, and when large quantities are involved, a truck foreman should be used to keep the trucking operation running efficiently.

A grade checker is used on the grade to spot the trucks for dumping. Although aggregate subbase may be moved about during the spreading and compacting operation, savings are achieved when the dumping is controlled to ensure that the tonnage dumped per linear foot of roadway is approximately equal to the theoretical amount required. The procedures used to determine the dumping rate are the same as those outlined in the following section for aggregate base except somewhat less precision is necessary.

6.2.2 Spreading and Compacting Aggregate Subbase

The unit costs to spread and compact aggregate subbase depend on the delivery rate for the material, and it is imperative that the delivery rate assumed on the estimate be achieved or bettered during the course of the work. When suppliers have quoted the material FOB the job, the contractor must satisfy himself that the required delivery rate will be realized with the loaders and trucks which the supplier proposes to use.

A typical string of equipment required to spread and compact aggregate subbase is:

1 motor grader
1 185-HP bulldozer pulling
1 vibratory steel roller (8000 pounds)
1 small rubber-tired tractor pulling a set of discs
1 or more water trucks

The labor required will be:

1 grade foreman
1 grade checker

The number of water trucks used in the estimate is based on past job experience or is estimated as shown in Section 6.1.7.1. Aggregate subbase is frequently quite dry, and it is essential to give careful consideration to the amount of water required.

The unit costs for spreading and compacting aggregate subbase are determined by assigning the appropriate hourly rental and wage rates to the equipment and labor shown in the preceding, and the resulting total hourly costs are divided by the estimated delivery rate for the typical job situation. The resulting base unit costs are increased (or the delivery rate used is decreased) for those sections which are more confined, and the amount of this adjustment will depend on the estimator's judgment. The unit costs for dumping, spreading, and compacting aggregate sub-base will be the same regardless of the number of layers required.

6.2.3 Making the Grade on Aggregate Subbase

It is necessary to make the grade only on the top layer of aggregate subbase since the approximate grades for the layers below are achieved, at no additional cost, by controlling the rate of dumping.

The base unit costs for making grade are indexed to the same equipment and crew used to make grade on earthwork (Section 6.1.12), and the production rate will be about 0.80 MSY per hour when the grade tolerance is ± 0.05 foot. Thus, based on $K_{MG-R} = 128$ and $K_{MG-L} = 61.78$, the base unit costs for equipment and labor will be:

$$\text{Equipment} \quad \frac{128}{0.80} = \$160 \text{ per MSY}$$

$$\text{Labor} \quad \frac{61.78}{0.80} = \$77.20 \text{ per MSY}$$

The preceding base unit costs assume quantities of 10 MSY or more and free access for the equipment.

6.2.4 Aggregate Base (AB)

Aggregate base is normally placed immediately above the aggregate subbase, and the thickness of the aggregate base section will vary from about 0.33 to 1 foot. The material is usually furnished by a commercial rock plant, although on large jobs it may be manufactured at the site.

Typical gradations for aggregate base are as follows:

	Percentage passing	
Sieve Sizes	$1\frac{1}{2}$ inches maximum	$\frac{3}{4}$ inches maximum
2 inches	100	—
$1\frac{1}{2}$ inches	87–100	—
1 inch	—	100

$\frac{3}{4}$ inch	45–90	87–100
No. 4	20–50	30–60
No. 30	6–29	5–35
No. 200	0–12	0–12

Specifications normally require aggregate base to contain sufficient moisture for compaction at the time it is delivered to the grade. This requirement is imposed in an attempt to prevent undue segregation during the spreading and compacting operation.

6.2.5 Spreading and Compacting Aggregate Base

When the required thickness is greater than 0.5 foot, the material is spread in two or more layers of approximately equal thickness, and the compacted thickness of each layer cannot exceed 0.5 foot. Each layer must be spread in one operation at a uniform rate per linear foot of roadway and must be a full traffic lane in width.

Aggregate base is usually delivered in end-dump trucks and dumped through a spreader box attached to the blade of a bulldozer. Some specifications allow the material to be dumped in windrows by bottom-dump trucks. In such cases, the material is spread with a 150-HP motor grader that has end wings attached to the blade at the required width.

In all cases, the tons of material spread or dumped per linear foot of lane must be carefully controlled so that the compacted thickness of each layer will equal its required thickness. Typical specifications will not allow any appreciable amount of material to be moved about or shifted during the grade-making operation.

The compacted density of the moist aggregate base will range from 1.80 to 2.00 tons per compacted cubic yard depending on the source of the material. When this density is not known, it is determined by laboratory tests.

The spreading or dumping rate for the material is determined as follows:

$$R = \frac{DWt}{27}$$

where

R = the spreading or dumping rate in tons per linear foot of lane
D = the density of the compacted moist material in tons per cubic yard
W = the width of the lane, or spread, in feet
t = the thickness of the layer to be spread in feet

Thus, assuming $D = 1.90$ tons per cubic yard, $W = 12$ feet, and $t = 0.5$ foot,

$$R = \frac{1.90 \times 12 \times 0.5}{27} = 0.422 \text{ tons per foot}$$

and a truck containing 23.9 tons of material would be spread or dumped in a distance of

$$\frac{23.9}{0.422} = 56.6 \text{ feet}$$

To control the spreading or dumping, the grade checker sets a temporary stake at this measured distance, and the distance required for the second load is measured from this stake, not from the end of the previous spread or windrow.

The base unit equipment cost for dumping, spreading, and compacting aggregate base when a spreader box is used depends on the delivery rate for the material and is indexed to the sum of the hourly rental rates for the following equipment:

1 150-HP bulldozer with spreader box
1 self-propelled pneumatic roller
1 motor grader
1 small (2000-gallon) water truck

The base unit cost for labor is indexed to the sum of the hourly wage rates for the following workmen:

1 grade foreman
1 grade checker

The preceding equipment and labor will dump, spread, and compact aggregate base at a maximum rate of 300 tons per hour. To achieve this rate, free access must be available. On narrow ramps the rate will drop to about 150 to 200 tons per hour. In very tight areas, the rate will be still lower. Six hundred tons of aggregate base were dumped, spread, and compacted in a strip 2.5 feet wide by 0.5 foot deep at a rate of 45 tons per hour. The unit cost for dumping, spreading, and compacting aggregate base is the same regardless of the number of layers required.

When the material is delivered in bottom-dump trucks, an additional grade checker is necessary to spot and dump the trucks, and chains are used to control the gate openings. The equipment and labor needed to spread and compact the material are the same as that listed in the preceding except that the bulldozer and spreader box are replaced with a 150-HP motor grader. Bottom-dumping aggregate base in windrows is advantageous when job conditions preclude fast delivery rates. The dumping is performed as a separate operation, and the increased cost due to the low delivery rate is small since only one workman is involved. When sufficient material has been delivered to the grade, the spreading and compacting operation is done at the 300-tons-per-hour rate.

The costs to deliver the material to the grade are based on the same considerations discussed previously in connection with aggregate subbase. About 5 minutes are required to spot and dump each truck whether through a spreader box or in windrows. When the material is loaded by a commercial plant, at least 5 minutes must be allowed for loading each truck, and in estimating haul times, consideration must be given to the time lost while the trucks move through the plant area.

As an example, assume the aggregate base will be plant loaded and that the one-way haul distances are 4 miles through a residential area and 20 miles on good county roads. Assume the trucks will carry an average of 23.5 tons per load, they will travel at 26 miles per hour in the residential area and 41 miles per hour on the county roads, and 10 minutes will be required to load and dump. Assuming that the stockpile is close to the public road, the cycle time for the trucks will be

$$\frac{8}{26} + \frac{40}{41} + \frac{10}{60} = 1.45 \text{ hours per load}$$

and each truck will deliver

$$\frac{23.5}{1.45} \times \frac{55}{60} = 14.9 \text{ tons per 55-minute hour}$$

Assuming a desired production rate of 300 tons per hour, the number of trucks required will be

$$\frac{300}{14.9} = 20.1$$

Twenty trucks will be used, and the hourly production will be $20 \times 14.9 = 298$ tons per hour. Assuming an hourly rental rate per truck of $30 per hour, the unit equipment cost to haul the material will be

$$\frac{20 \times 30}{298} = \$2.01 \text{ per ton}$$

The foregoing calculation assumes a large job where two or more weeks of hauling are involved. In such cases it pays to bring in trucks, if necessary, from outlying areas. If the job were small and only a day or two of hauling were required, the number of trucks used would be based on the estimator's judgment of the number that would be available, locally, for the short duration of the work.

In such a case, it might be concluded that 10 trucks would be available; the effective production rate would then be

$$10 \times 14.9 = 149 \text{ tons per hour}$$

and the costs to spread and compact the material would be increased.

On the large job, the haul labor would consist of a load counter and a truck foreman. On the short job, only a load counter would be used, and the job superintendent would oversee the trucking operation.

6.2.6 Making the Grade on Aggregate Base

It is necessary to make the grade only on the top layer of aggregate base. The approximate grades for the layers below are achieved at no additional cost by controlling the rate of spreading and dumping.

The base unit costs for making grade are indexed to the same equipment and crew used to make grade on earthwork and on aggregate subbase, and the production rate will be about 0.87 MSY per hour when the grade tolerance is ± 0.05 foot, when the quantity involved is 3 MSY or more, and when there is free access for the equipment.

6.2.7 Prime Coats

The specifications may require applying a prime coat of liquid asphalt to the finished surface of the aggregate base course. The asphalt is paid for under a separate job item on a price-per-ton basis.

SC-250 liquid asphalt is commonly used for prime coats. The material is sprayed from insulated and metered distributor trucks at temperatures of 140 to 225 degrees Fahrenheit, and the material is applied at uniform rates of about 0.25 gallon per square yard. Typical distributor trucks carry a maximum of about 11 tons of asphalt and must carry a minimum of about 4 tons to prevent undue cooling of the material in transit.

Liquid asphalt is quoted, by the suppliers, at a price per ton which includes furnishing the material and applying it at the specified rate. Although the contractor incurs no additional costs in connection with the item, the quotations must be carefully analyzed since the actual cost per ton may be considerably higher than the base price quoted. This increased cost results from several factors which will vary with different suppliers but which may include such items as minimum load charges, a flat handling charge when unused material is returned, a reduced price credited for returned material, and freight charges which may be on a per load or on a per ton hauled basis.

AVERAGE WEIGHTS AND VOLUMES
OF LIQUID ASPHALT

Grade of Liquid Asphalt	Gallons per Ton at 60°F
70	253
250	249
800	245
3000	241

The volume–temperature relationship for SC, MC, and RC liquid asphalts, grades 250 through 3000, is

$$V_T = V_{60} \times F$$

where

V_T = volume at T degrees Fahrenheit

V_{60} = volume at 60°F

F = a multiplier which equals 1.0000 for $T = 60$, increases 0.00035 per degree over 60, and decreases by the same amount per degree under 60

6.2.7.1 PENETRATION TREATMENT. Seventeen tons of SC-250 penetration treatment was sprayed on a roadside ditch at a rate of 0.25 gallons per square yard. The liquid asphalt was applied with a hand spray attached to an insulated distributor truck at an average rate of 0.57 tons per hour.

6.2.8 Cement-treated Base

Cement-treated base (CTB), when used, is the top base course in the structural section for the roadway. The material may be plant mixed or road mixed, and the method of mixing will be specified. Cement-treated base courses vary from about 0.50 to 1 foot in thickness and are spread in layers having a maximum compacted thickness of 0.50 foot. The material is compacted to a relative compaction of 95 percent, and the compacted density will range from 1.80 to 2.00 tons per CY.

Large general contractors specializing in roadwork frequently place CTB as well as portland cement concrete pavement with their own crews and equipment. Much of this work is, however, done on a subcontract basis. Electronically controlled paving machines are used to spread and compact the material, which may be spread up to four lanes in width. Grade tolerance problems make it desirable to have the CTB course and the paving placed by the same subcontractor.

6.2.9 Plant-mixed CTB

Plant mixing, when required, is usually done by separate subcontractors who set up portable mixing plants at the job site and who will mix from 150 to 450 tons of CTB per hour. The cement, aggregates, and water are supplied to the plant by the general contractor, who may also deliver the mixed material to the grade.

The cement companies quote the cement, loaded into silos, at the mixing plant. The cement is paid for by the general contractor, and an allowance of about 5 percent is made for waste.

6.2.10 Road-mixed CTB

The aggregates for road-mixed CTB are normally delivered to the grade by the general contractor in bottom-dump trucks. The material is dumped at a carefully controlled rate per linear foot of windrow, and the procedures necessary to determine the dumping rate are the same as those described in connection with aggregate base. About 5 minutes are required to dump each truck, and a grade checker is needed to spot the trucks and control the dumping operation.

The windrows are then shaped to a uniform cross section with a sizing device which may be attached to a motor grader, and such shaping is normally done by the CTB subcontractor. Not more than 4 hours before mixing, the cement is metered onto the windrows from trucks furnished by the cement supplier; this operation does not entail any additional labor cost to the contractor, but 5 percent is allowed for waste cement.

The windrowed materials are then picked up and mixed by a road-mixing machine, and the water is supplied to the machine by the general contractor, who furnishes one or more water trucks for this purpose; from 6 to 7 percent, by weight, of water will be required.

6.2.11   Job Pits for CTB Aggregates

The aggregates for CTB are often produced from pits located near the job site. Oversize material must normally be removed, and belt loaders are used to load and separate the material.

A 385-HP bulldozer together with a 48-inch belt loader equipped with a 4 × 9 foot vibrating single-deck screen loaded trucks at a rate of 475 tons per 60-minute hour. Approximately 6.5 percent of the pit material was scalped off, and the total production for the bulldozer and belt was 508 tons per hour. The 24-ton trucks averaged 23.7 tons per load, and a small wheel loader was used to clear the scalped material from the loading area.

6.2.12   Blending CTB Aggregates

The pits used for the job production of CTB aggregates are frequently deficient in rock, which is the material retained on the No. 4 sieve. In such cases, additional rock must be imported or some of the pit sand (material passing the No. 4 sieve) must be wasted. Separating the sand to be wasted is impractical for a job plant, so additional rock is imported.

A typical grading specification for CTB aggregates is shown on line 1 of Table 6.5. Line 2 shows the gradation of an assumed pit sample; 7.0 percent of the pit material is oversize and must be wasted.

Line 3 shows the gradation of that fraction of the pit sample which passed the 1-inch sieve, and line 4 shows the gradation of the rock to be imported. The imported rock will be blended with the scalped pit material so that 70.0 percent of the combined product will pass the No. 4 sieve. This percentage will furnish a leeway, or safety factor, of 5 percent with respect to the specifications.

Since 80.8 percent of the scalped pit product is sand (line 3, column 5), the percentage of rock is $100.0 - 80.8 = 19.2$ percent. Since 80.8 percent of the scalped pit product must equal 70.0 percent of the combined product, the following proportion is set up:

$$\frac{Rc}{19.2} = \frac{70.0}{80.8}$$

TABLE 6.5   Material Grading

Material	% Used	2"	1"	$\frac{3}{4}$"	No. 4	No. 30	No. 200
				Sieves—% Passing			
(1) CTB specification	—		100	87–100	35–75	7–45	0–19
(2) Typical pit sample		100	93.0	—	—	—	—
(3) Scalped pit sample	—		100	98.1	80.8	36.0	9.2
(4) Imported rock	—		100	100	0	0	0
(5) Scalped pit	86.6	—		86.6	85.0	70.0	31.2	8.0
(6) Imported rock	13.4	—		13.4	13.4	—	—	—
(7) Combined product	100	—		100	98.4	70.0	31.2	8.0

where

   Rc = the percentage of rock in the combined product

and

$$Rc = 19.2 \times \frac{70.0}{80.8} = 16.6 \text{ percent}$$

These data are tabulated as follows:

Pit Material	Percentage of Pit Product	Percentage of Combined Product
Sand	80.8	70.0
Rock	19.2	16.6
	100.0	86.6

The percentage of imported rock required is $100.0 - 86.6 = 13.4$ percent of the combined product.

The percentages of material in the combined product are shown on lines 5, 6, and 7 of Table 6.5. The percentages of scalped pit material shown on line 5 equal 86.6 percent of the percentages shown on line 3. Similarly, the percentages of imported rock shown on line 6 equal 13.4 percent of the percentages shown on line 4.

The combined product, shown on line 7, is the sum of lines 5 and 6 and meets the specified gradation.

When CTB is plant mixed, the imported rock is delivered to a separate stockpile at the plant, and the blending entails no additional cost. When the CTB is road mixed, the pit material is dumped in a controlled windrow from bottom-dump trucks, and the imported rock is dumped in a second controlled windrow, directly on top of the first one. The dry aggregates are then mixed with one or more passes of the road-mixing machine, the blended windrow is sized, and it is then ready for the cement.

Assume that a 4.5-cubic-foot windrow containing 0.30 tons of aggregates per linear foot is required and that the materials shown on lines 5 and 6 of Table 6.5 will be used. The amount of pit material windrowed is

$$0.30 \times 0.866 = 0.260 \text{ tons per linear foot}$$

and the amount of imported rock required is

$$0.30 \times 0.134 = 0.040 \text{ tons per linear foot}$$

Assuming the material is delivered in 24-ton loads, the pit material will be dumped in a measured distance of

$$\frac{24}{0.26} = 92.3 \text{ feet}$$

and the imported rock trucks will be dumped in a measured distance of

$$\frac{24}{0.04} = 600 \text{ feet}$$

In this case, a grade checker will dump the pit trucks in about 5 minutes but will require about 15 minutes to dump the imported rock trucks.

6.2.13 Curing CTB

The completed CTB is covered with a bituminous curing seal. MC-250 liquid asphalt is used and is applied at a rate of from 0.15 to 0.25 gallons per square yard of surface. The procedure is identical to that described for placing the prime coat on aggregate base and entails no additional labor costs for the general contractor.

6.2.14 Portland Cement Concrete Pavement

Portland cement concrete pavement is often placed by subcontractors. Sometimes the work is done by paving firms that do nothing but this type of work, and in other cases the pavement is placed, on a subcontract basis, by large general contractors who specialize in the construction of roads and highways.

The job batch plants and paving machines required are expensive, highly specialized, and not available for rental (Figure 6.9). Due to the high cost, this equipment is owned only by contractors who do large amounts of portland cement concrete paving. Highway contractors do this work on a subcontract basis in an effort to keep their expensive equipment working.

Figure 6.9. Slipform paving machine (Owl Slipform Concrete Co.).

The general contractor furnishes the site for the plant itself, and from 5 to 7 acres are required. He also furnishes the water supply; about 350 gallons per minute are needed assuming a production rate of 500 CY per hour. The general contractor usually furnishes the aggregates delivered to the job plant and will, in all cases, furnish and maintain the necessary job haul roads. The cement is normally set up as a separate job item and is sometimes furnished by the general contractor.

Cement companies quote the cement loaded into job silos and normally furnish the necessary silos at no additional cost. The general contractor incurs no labor nor equipment costs in connection with the cement item, but an allowance of 5 percent is made for waste cement. The yield for the mix design used must be carefully checked since large quantities of cement are required, and the yield should be rechecked as the work progresses.

The aggregates are delivered in bottom-dump trucks, dumped into receiving hoppers, and loaded onto stockpiles by conveyors. The equipment required for unloading is furnished and installed by the paving subcontractor. About 1.07 tons of rock and 0.66 ton of sand are required per CY of concrete, including the allowance for waste.

Assuming a slipform paver is used and the batch plant is equipped with two 8-CY mixers, the maximum production rate will be about 500 CY per hour. The mixed concrete is delivered to the grade in end-dump trucks, which are rented by the paving contractor.

Assuming a production rate of 500 CY per hour, about 36 loads of aggregates and 62 loads of mixed concrete are delivered per hour. At the same time, on an average job, aggregates for base courses are arriving, and the resulting truck traffic on the job site prevents the production rate from being increased efficiently even though larger mixers might be used.

The aggregates are normally obtained from commercial rock plants, and in view of the quantities involved, the general contractor should load the trucks. In addition, a truck foreman and a load counter should be used to keep the operation running efficiently. It is essential that the job haul roads be well maintained since large quantities of materials will be hauled, and the aggregate trucks travel at speeds up to 40 miles per hour in many instances. There are no labor costs in connection with haul road maintenance except for the costs of flagmen, if required.

6.2.15 Asphalt Pavements

Asphalt concrete pavement and plant-mixed surfacing are typically placed by sub-contractors, and their effective production rate ranges from 1800 to 2000 tons per day when working in unconfined areas. The average in-place density of both types of asphalt paving is about 2 tons per CY. The general contractor will incur no expenses in connection with the asphalt paving item except for specific incidental items of work which may be excluded in the quotation by the subcontractor.

Plant-mixed surfacing is sometimes used to form spillways on embankment slopes. Based on $K_L = 13.48$ and $K_{BH} = 25$, areas averaging 4 SY each were graded at unit costs of $6.00 per SY for labor and $1.50 per SY for equipment. A $1\frac{1}{2}$-inch-thick layer of plant-mixed surfacing was placed and compacted by hand on these areas at a unit labor cost of $5.20 per SY.

Based on $K_L = 13.48$ and $K_{MR} = 74$, plant-mixed surfacing was placed over pipe trenches at unit costs of $12.50 per ton for labor and $5.10 per ton for equipment, where K_{MR} = the sum of the hourly rental rates for one motor grader plus one self-propelled steel roller. About 18 tons of material were placed at each trench.

Based on $K_L = 13.48$ and $K_{BH} = 25$, manhole covers were raised and adjusted to grade, after paving, at unit costs of $124 each for labor and $3.20 each for equipment, and these prices included placing the necessary concrete manhole rings and the cost to hand-place surfacing at the raised manhole cover on completion.

6.3   CLEARING AND GRUBBING: FINISHING ROADWAY

Clearing and grubbing is the first item done, and finishing the roadway is the last. In both cases, road-building equipment is used to do the work.

6.3.1 Clearing and Grubbing

Clearing and grubbing is usually bid at a lump sum price. Since the item is performed early and the quantity is fixed, clearing and grubbing is normally bid at a price considerably greater than the bare cost for doing the work. In many instances, the price bid for clearing and grubbing will be about 10 percent of the total job price.

Based on $K_L = 13.48$ and a rental rate of $39 per hour for a motor grader, areas ranging from 2000 to 5000 MSY covered with sagebrush and light mesquite have been cleared at average unit costs of about $2.50 per MSY for labor and about $7.60 per MSY for equipment. The unit cost for labor represents the cost to burn the brush.

Trees greater than about 3 inches in diameter are cleared with bulldozers having a minimum of 250 HP, and when the trees are greater than about 12 inches in diameter, a 385-HP bulldozer with a machine-mounted ripper should be used.

The number of trees to be removed is based on an actual count for small areas and on the counted number of trees in a typical acre when the areas are large. The time required to uproot trees is about as follows:

Diameter (inches)	Time (minutes)
4–8	3
12	6
24	15
30	20

The roots of the larger trees are cut with the ripper, and the trees are then pushed over. The preceding times include the time required for ripping, uprooting, and filling the resulting holes.

When large numbers of trees must be removed, subcontractors are normally used. Such subcontractors have the necessary equipment, skilled crews, and supervision and often have a market for the trees they remove.

6.3.2   Finishing Roadway

Upon completion of all construction operations, the roadway must be cleaned up finished as specified. Typical work required is:

1. Trim and finish all embankment and cut slopes.
2. Trim, shape, and recompact the shoulders, which have been disturbed during paving and slope trimming.
3. Clean out drainage ditches and pipes as required.
4. Remove all weeds.
5. Sweep paved areas.
6. Clean up and remove all debris, and leave the site in a neat and orderly condition.

The finishing roadway item is usually bid at a lump sum price, and since the item is the last one completed, the price bid should be no greater than the estimated bare cost for doing the work. A review of the bid prices for the finishing roadway item indicates that it is frequently bid at about half of this amount.

6.3.3 Trimming Slopes

Slopes up to 12 feet high are trimmed with a motor grader; one grade checker and one laborer will be required. About 100 stations of slope on one side of the roadway can be trimmed per 8-hour day.

Slopes up to 30 feet in height are trimmed by dragging anchor or dredge chain with two pieces of equipment, and the chain used will weigh from 45 to 60 pounds per linear foot. A motor grader or a wheel loader is used on the grade, and a bulldozer is used on the natural ground. One grade checker and about four laborers will be required, and about 10 stations of slope on one side of the roadway can be trimmed per hour.

The cost to trim embankment slopes at bridge abutments depends on the accuracy of the slope stakes used and on the amount of care exercised in placing the fill. Embankment construction at bridge abutments should be carefully supervised since in most cases $1\frac{1}{2}:1$ slopes under the bridge are intersecting with $2:1$ slopes at the sides. The slope stakes at the intersections must be set at the proper radius for each stake, and the grade checker must check the grade during construction on the radial to the stake being used.

The cost to trim the abutment slopes for a highway overcrossing having a width of about 36 feet may be estimated by allowing from 8 to 16 hours for a 270-HP bulldozer, one laborer, and one grade checker per bridge. Skewed bridges with ramps require more time than square bridges with no ramps.

6.3.4 Trimming Shoulders

One motor grader, one self-propelled pneumatic roller, one-half water truck, and one grade checker are required. About 30 stations of shoulder up to 8 feet in width will be backed up and trimmed per hour.

6.3.5 Cleaning Ditches and Culverts

The estimated cost for this work depends on the estimator's judgment in view of the particular job conditions. On most well-run jobs where the downstream drainage has been completed, very little work will be required. An average allowance of about 2 hours of laborer time per culvert and per drainage inlet will cover the cost of whatever work may be required.

6.3.6 Remove Weeds

This work is done with a motor grader, and no additional labor is required where the weeds are small. The production rate will be about 10 MSY per hour.

6.3.7 Sweep Paved Areas

A power road broom is used and no additional labor is required. The production rate will be about 12 MSY per hour.

6.3.8 Cleanup

The incidental cleanup costs represent a substantial portion of the total cost for the finishing roadway item, and in general, these costs cannot be predicted with much accuracy at the time the estimate is being prepared. Based on $K_L = 13.48$ and $K_M = 39$ (where K_M = the hourly rental rate for a motor grader), the incidental costs have averaged about $11 per station for labor and $6 per station for equipment on freeway jobs in rural areas where 1000 or more stations were involved.

6.3.9 Overall Unit Costs

The estimated cost for finishing the raodway is often based on cost records showing the total labor and equipment costs for similar jobs. Such total costs are converted to unit costs per station of roadway. The stations used include the stations of all ramps and frontage roads; when there is a divided highway, the stations of both sides are included in the total. Thus, if a given project consists of 150 stations of divided highway, 70 stations of ramps, and 170 stations of frontage roads, the unit costs would be based on $150 + 150 + 70 + 170 = 540$ stations of roadway.

Based on $K_L = 13.48$ and $K_M = 39$ (where K_M = the hourly rental rate for a motor grader), the unit costs realized for the entire finishing roadway item have averaged as follows:

Freeways in urban areas, 22 to 320 stations in total length:

Labor	$33 per station
Equipment	$27 per station

Freeways in rural areas 1000 or more stations in total length:

Labor	$21 per station
Equipment	$20 per station

City streets, with curbs and sidewalks, 50 stations in total length:

Labor	$39 per station
Equipment	$14 per station

Small jobs in rural areas, 2 to 3 stations in total length:

Labor	$20 per station
Equipment	$60 per station.

6.4 DRAIN PIPES

Culvert pipe and pipework relating to other surface drainage are normally installed by the general contractor. This work must usually be done in segments spread over the life of the job, and subcontracting is often not feasible due to scheduling difficulties and excessive on-and-off costs.

Sewer lines and water piping are installed by subcontractors. When culvert pipes must be jacked under roadways, this work is also done by specialists on a subcontract basis. In most cases the general contractor will furnish the jacking pit, which will be about 4 feet wider than the outside pipe diameter and about 4 feet longer than the pipe joint length and will have a vertical back face.

Pipe excavation and backfill are normally carried under separate job items, as discussed in Chapter 5. The unit costs herein for installing pipes include the costs for fine grading but do not include the costs for excavation or for backfill.

6.4.1 Corrugated Steel Pipe and Pipe Arches

The weights of typical riveted corrugated steel pipes are given in Tables 6.6 and 6.7. Table 6.6 covers pipes with $2\frac{2}{3}$-inch-pitch by $\frac{1}{2}$-inch-deep corrugations; Table 6.7 covers pipes with 3-inch-pitch by 1-inch-deep corrugations. The standard joint length is 20 feet, and shorter pieces are prefabricated to order, as required. Typically, the suppliers take off the lengths required and quote the material FOB the job.

TABLE 6.6   **Approximate Weights for Corrugated, 2 2/3 × 1/2-inch Steel Pipe**[a]

Inside Diameter (inches)	Specified Thickness (inches)	Approximate Pounds Per Linear Foot			
		Galvanized	Full Coated	Full Coated and Invert Paved	Full Coated and Full Paved
12	0.064	10	12	15	
	0.079	12	14	17	
15	0.064	12	15	18	
	0.079	15	18	21	
18	0.064	15	19	22	
	0.079	18	22	25	
	0.109	24	28	31	
21	0.064	17	21	26	
	0.079	21	25	30	
	0.109	29	33	38	
24	0.064	19	24	30	45
	0.079	24	29	35	50
	0.109	33	38	44	60
30	0.064	24	30	36	55
	0.079	30	36	42	60
	0.109	41	47	53	75

TABLE 6.6   Continued

Inside Diameter (inches)	Specified Thickness (inches)	Approximate Pounds Per Linear Foot			
		Galvanized	Full Coated	Full Coated and Invert Paved	Full Coated and Full Paved
36	0.064	29	36	44	65
	0.079	36	43	51	75
	0.109	49	56	64	90
	0.138	62	69	77	100
42	0.064	34	42	51	—
	0.079	42	50	59	85
	0.109	57	65	74	105
	0.138	72	80	89	115
48	0.064	38	48	57	—
	0.079	48	58	67	95
	0.109	65	75	84	120
	0.138	82	92	101	130
	0.168	100	110	119	155
54	0.079	54	65	76	105
	0.109	73	84	95	130
	0.138	92	103	114	155
	0.168	112	123	134	175
60	0.079	60	71	85	—
	0.109	81	92	106	140
	0.138	103	114	128	180
	0.168	124	135	149	190
66	0.079	65	77	93	
	0.109	89	101	117	160
	0.138	113	125	141	180
	0.168	137	149	165	210
72	0.109	98	112	129	170
	0.138	123	137	154	210
	0.168	149	163	180	240
78	0.109	105	121	138	200
	0.138	133	149	166	230
	0.168	161	177	194	260
84	0.138	144	161	179	240
	0.168	173	190	208	270
90	0.138	154	172	192	
	0.168	186	204	224	
96	0.168	198	217	239	

[a]Lock seam construction only; weights will vary with other fabrication methods.

Source: Armco Inc., Metal Products Division.

The weight per linear foot for a pipe arch is equal to the weight of a pipe section having the same periphery. The diameter of the equivalent pipe section is approximately equal to the average of the two pipe arch dimensions multiplied by 1.03.

Corrugated steel pipe is designated by wall thickness or by gage of the metal used. The relationship between wall thickness and gage is:

Wall Thickness (inches)	Gage
0.064	16
0.079	14
0.109	12
0.138	10
0.168	8

Since the weight of a given joint will vary by more than 100 percent depending on the gage, corrugations, and coatings, the costs to install the material are based on the weight per linear foot of pipe or of pipe arch.

When corrugated steel pipes are installed, it is imperative that the outside circumferential laps be upstream, and longitudinal laps should not be located in the invert.

TABLE 6.7 **Approximate Weights for Corrugated, 3 × 1-inch Steel Pipe**[a]

		Approximate Pounds per Linear Foot			
Inside Diameter (inches)	Specified Thickness (inches)	Galvanized	Full Coated	Full Coated and Invert Paved	Full Coated and Full Paved
36	0.064	33	44	56	92
	0.079	41	52	64	100
	0.109	56	67	79	115
	0.138	71	82	94	130
	0.168	87	98	110	146
42	0.064	39	52	66	107
	0.079	47	60	74	116
	0.109	65	78	92	134
	0.138	83	96	110	152
	0.168	100	113	127	169
48	0.064	44	59	75	123
	0.079	54	69	85	132
	0.109	74	89	105	152
	0.138	95	110	126	174
	0.168	115	130	146	194
54	0.064	50	66	84	138
	0.079	61	77	95	149
	0.109	83	100	118	171
	0.138	106	123	140	194
	0.168	129	146	163	217
60	0.064	55	73	93	153
	0.079	67	86	105	165
	0.109	92	110	130	190
	0.138	118	136	156	216
	0.168	143	161	181	241
66	0.064	60	80	102	168
	0.079	74	94	116	181
	0.109	101	121	143	208
	0.138	129	149	171	236
	0.168	157	177	199	264

TABLE 6.7 Continued

Inside Diameter (inches)	Specified Thickness (inches)	Approximate Pounds per Linear Foot			
		Galvanized	Full Coated	Full Coated and Invert Paved	Full Coated and Full Paved
72	0.064	66	88	111	183
	0.079	81	102	126	197
	0.109	110	132	156	227
	0.138	140	162	186	257
	0.168	171	193	217	288
78	0.064	71	95	121	198
	0.079	87	111	137	214
	0.109	119	143	169	246
	0.138	152	176	202	279
	0.168	185	209	235	312
84	0.064	77	102	130	213
	0.079	94	119	147	230
	0.109	128	154	182	264
	0.138	164	189	217	300
	0.168	199	224	253	335
90	0.064	82	109	140	228
	0.079	100	127	158	246
	0.109	137	164	195	283
	0.138	175	202	233	321
	0.168	213	240	271	359
96	0.064	87	116	149	242
	0.079	107	136	169	262
	0.109	147	176	209	302
	0.138	188	217	250	343
	0.168	228	257	290	383
102	0.064	93	124	158	258
	0.079	114	145	179	279
	0.109	155	186	220	320
	0.138	198	229	263	363
	0.168	241	272	306	406
108	0.079	120	153	188	295
	0.109	165	198	233	340
	0.138	211	244	279	386
	0.168	256	289	324	431
114	0.079	127	162	199	312
	0.109	174	209	246	359
	0.138	222	257	294	407
	0.168	271	306	343	456
120	0.109	183	220	259	378
	0.138	234	271	310	429
	0.168	284	321	360	479

[a]Lock seam construction only; weights will vary with other fabrication methods.

Source: Armco Inc., Metal Products Division.

Average unit costs for installing plain galvanized corrugated steel pipe or pipe arches are given by the following empirical formulas, which are based on cost data from many typical installations. These unit costs include the costs of unloading the material near the place of installation.

Equipment is not required to install pipes weighing up to about 30 pounds per linear foot, and

$$C_{LS} = 1.23 + 0.048W \qquad (6.11)$$

where

C_{LS} = labor cost in dollars per linear foot, indexed to K_L = 13.48, for pipes weighing from 8 to 30 pounds per linear foot

W = the weight of the pipe or of the pipe arch in pounds per linear foot

The average unit costs to install the heavier pipes or pipe arches are about

$$C_{LL} = 0.55 + 0.040W \qquad (6.12)$$

and

$$C_{RL} = 0.34 + 0.014W \qquad (6.13)$$

where

C_{LL} = labor cost in dollars per linear foot, indexed to K_L = 13.48, for pipes weighing from 30 to 200 pounds per linear foot

C_{RL} = the unit rental cost corresponding to C_{LL}, indexed to K_{BH} = 25

W = the weight of the pipe or of the pipe arch in pounds per linear foot

When bituminous coated pipe is specified, the unit weights for the coated pipe should be used in the preceding formulas, and the resulting unit costs should be increased by 25 percent.

Assume the following conditions:

1. Plain galvanized corrugated steel pipe 36 inches in diameter by 12 gage, weight 49 pounds per linear foot.
2. Hourly wage rate for a laborer $15.00.
3. Hourly rental rate for a small rubber-tired backhoe $29.00.

Then the estimated unit costs to install the material will be

$$C_{LL} = (0.55 + 0.040 \times 49) \, \frac{15.00}{13.48} = \$2.80$$

and

$$C_{LR} = (0.34 + 0.014 \times 49)\frac{29.00}{25} = \$1.20$$

Corrugated steel culverts normally have concrete headwalls at each end (minor structures—Section 8.1.71) or galvanized steel flared end sections.

ESTIMATED UNIT COSTS TO INSTALL PLAIN GALVANIZED STEEL FLARED END SECTIONS

Pipe diameter (inches)	Labor cost ($K_L = 13.48$)	Rental cost ($K_{BH} = 25$)	Unit of Measure
12–15	$ 40	—	Each
18	50	—	Each
24	60	—	Each
30	75	—	Each
36	60	$25	Each
42	70	25	Each
48	80	35	Each
60	110	50	Each

When bituminous coated end sections are specified, the preceding unit costs should be increased by about 25 percent.

Galvanized steel entrance tapers are normally used with 8- and 12-inch corrugated steel pipe down drains. Indexed to $K_L = 13.48$, the labor cost to install these is about $48 each for the 8-inch tapers and about $60 each for the 12-inch tapers. When bituminous coated tapers are specified, these costs should be increased by about 25 percent.

The preceding unit costs for installing flared end sections and entrance tapers include the costs for fine grading and all incidental excavation and backfill.

6.4.2 Salvaging Corrugated Steel Pipe

The unit labor cost to salvage corrugated steel pipe, whether plain galvanized or coated, will run about 2 to 2.5 times the unit labor cost for installing the plain galvanized material. The unit cost for equipment will be the same as the installation cost for plain galvanized pipe.

6.4.3 Corrugated Steel Multiplate Structures

Multiplate structures are used when large water areas are required. Prefabricated corrugated steel plates together with the necessary bolts are supplied by the manufacturers and are assembled in place, on the job. Multiplate pipes range from about

5 to 15 feet in diameter, multiplate pipearches have spans ranging from about 6 to 17 feet, and multiplate arch spans range from about 6 to 30 feet.

Multiplate structures are erected by subcontractors specializing in this work, and the names of these subcontractors, for a given area, are obtained from the material suppliers.

6.4.4 Rubber-Gasketed Reinforced Concrete Pipe

The weights per linear foot, the wall thicknesses, and the standard joint lengths for reinforced concrete pipes vary with the different manufacturers. Typical dimensions and weights for Class III rubber-gasketed reinforced concrete pipes are shown in Table 6.8.

Unit costs for installing rubber-gasketed reinforced concrete pipe having 11-foot joint lengths are shown in Table 6.9. It is assumed the pipe is unloaded by the supplier, at no additional cost, near the place of installation.

The unit costs shown in Table 6.9 assume average conditions and trench depths such that shoring is not required. When pipes are installed in shored trenches, the unit costs should be increased about 25 percent. The unit costs include the costs for fine grading the trench but do not include the costs for excavation or backfill, which are estimated as separate items.

Although the unit costs for rental are indexed to the hourly rental rate for a 35-ton truck crane, a smaller crane is satisfactory for the smaller pipe sizes and a larger crane may be necessary for pipes over 5 feet in diameter. In all cases, the estimated unit cost for rental should be based on the hourly rental rate for a crane that can easily handle the pipe under the anticipated job conditions.

6.4.5 Hydrostatic Tests

Hydrostatic testing is normally required when rubber-gasketed reinforced concrete pipe is installed. After the pipe has been backfilled to a minimum depth of about

TABLE 6.8   Rubber-gasketed Reinforced Concrete Pipe (Class III)

Inside Diameter (inches)	Wall Thickness (inches)	Weight (pounds per linear foot)	Weight (tons per 11-foot joint)
12	2	105	0.58
18	2.25	170	0.94
24	2.50	255	1.40
30	2.75	350	1.93
36	3.125	475	2.61
42	3.75	675	3.71
48	4.125	850	4.68
54	4.5	1075	5.91
60	6	1375	7.56
66	6.5	1620	8.91
72	7	2000	11.0

TABLE 6.9 Unit Costs for Installing Rubber-gasketed Reinforced Concrete Pipe (11-foot joint lengths)

Pipe Diameter (inches)	Unit Costs		Unit of Measure
	Labor (K_L = 13.48)	Rental[a] (K_{C_r} = 75)	
12	$1.30	$ 1.10	LF
18	1.30	1.30	LF
24	1.30	1.60	LF
30	1.40	1.90	LF
36	1.50	2.30	LF
42	1.70	2.80	LF
48	2.10	3.40	LF
54	3.50	5.20	LF
60	4.90	7.20	LF
66	6.40	9.20	LF
72	7.80	12.00	LF

[a]The costs for moving the crane on and off the job are additional.

2 feet, the test will typically be conducted over a 24-hour period with a 10-foot hydrostatic head on the highest point of the pipe. All obvious leaks must be stopped, and no leakage over the amount permitted by the specifications is allowed. The amount of leakage permitted is about 0.60 gallons per inch of inside diameter per 100 feet of pipe per hour.

For testing purposes, the ends of the pipe are normally sealed through the use of special end joints, which are furnished with concrete plugs in place. These test joints are supplied on a loan basis by the pipe manufacturer, and except for freight charges, no costs are involved. Nozzles are provided in the plugs for connecting the stand pipe and the supply hose and for draining the line upon completion of the test.

The operations involved in making a hydrostatic test are:

1. The test joints must be installed and later removed.
2. The ends of the test joints must be strutted or backfilled to prevent the pipe joints from spreading under the hydrostatic head.
3. The line must be filled with water.
4. Any leaks which develop must be stopped.

Although testing costs are subject to considerable variation, typical costs when test joints are used are shown in Table 6.10.

In most cases, a water truck is used to fill the pipe; under average job conditions, the water truck averages about three loads per hour.

The amount of water required to fill a given pipe is

$$M = 0.00408D^2$$

TABLE 6.10 Unit Costs for Hydrostatic Testing[a]

Pipe Diameter (inches)	Unit Costs per Test		Hundred Gallons per 100 LF
	Labor ($K_L = 13.48$)	Rental ($K_{Cr} = 75$)	
12	$130	$140	0.59
24	160	150	2.35
36	200	160	5.29
48	270	220	9.40
60	380	370	14.7
72	520	550	21.2

[a]The costs for furnishing the required water and for filling the section to be tested are additional.

where

M = thousands of gallons per 100 linear feet of pipe

D = pipe diameter in inches

Under normal conditions, there is no additional labor cost to fill the section with water.

It is sometimes necessary to seal the ends of a pipe line with timber bulkheads when test joints are not available. Bulkheads for a 60-inch pipe will be installed and removed for a labor cost of about $420 per bulkhead indexed to $K_{6+2} = 1000$. This cost assumes the length of the pipe line to be such that the bulkheads may be tied together with steel rods.

6.4.6 Tongue-and-Groove Reinforced Concrete Pipe

The weights per linear foot and the standard joint lengths for tongue-and-groove pipe vary with the different manufacturers. Typical weights for Class III tongue-and-groove pipes are shown in Table 6.11.

TABLE 6.11 Tongue-and-Groove Reinforced Concrete Pipe (Class III)

Inside Diameter (inches)	Wall Thickness (inches)	Weight (pounds per linear foot)	Weight (tons per 8-foot joint)
12	2	100	0.40
18	2.25	160	0.64
24	2.50	230	0.92
30	2.75	315	1.26
36	3.125	425	1.70
42	3.875	625	2.50
48	4.25	775	3.10
54	4.625	950	3.80
60	5.25	1200	4.80
66	5.75	1450	5.80
72	6.25	1710	6.84
84	7	2590	10.4

Unit costs for installing tongue-and-groove reinforced concrete pipe having 8-foot joint lengths are shown in Table 6.12. It is assumed the pipe is unloaded by the supplier near the place of installation at no additional cost.

The comments made regarding Table 6.9 apply here also.

Tongue-and-groove pipe joints are sealed with a cement mortar band. The average quantities of cement required are shown in Table 6.12 and assume a mortar consisting of one part cement to two parts sand. About 0.12 ton of sand are required per sack of cement, but since the sizes of the mortar bands are not specified, the amounts of material required vary.

The unit costs shown in Table 6.12 include the costs for the following operations:

1. Fine grading the trench.
2. Setting the pipe.
3. Installing and curing the mortar bands.

Although the equipment rental costs relate solely to the setting of the pipe, the labor cost breaks down about as follows:

1. Fine grading	22 percent
2. Setting	21 percent
3. Installing mortar bands	57 percent
	100 percent

6.4.7 Concrete Flared End Sections

The dimensions and weights of reinforced concrete flared end sections vary with different manufacturers. Typically, a small cast-in-place concrete cutoff wall 8

TABLE 6.12 Unit Costs for Installing Tongue-and-Grove Reinforced Concrete Pipe (8-foot joint lengths)

Pipe Diameter (inches)	Unit Costs		Cement (sacks per joint)	Unit of Measure
	Labor ($K_L = 13.48$)	Rental[a] ($K_{Cr} = 75$)		
12	$ 1.10	$ 0.80	0.05	LF
18	1.60	1.10	0.09	LF
24	2.00	1.40	0.13	LF
30	2.60	1.80	0.18	LF
36	3.30	2.30	0.23	LF
42	4.20	3.00	0.32	LF
48	5.20	3.90	0.41	LF
54	6.30	5.00	0.50	LF
60	7.80	6.50	0.59	LF
66	9.50	8.30	0.74	LF
72	11.50	10.60	0.89	LF
84	17.20	17.60	1.0	LF

[a]The costs for moving the crane on and off the job are additional.

TABLE 6.13 Average Unit Costs to Install Reinforced Concrete Flared End Sections

Pipe Diameter	Unit Costs		Cutoff Wall (CY)	Unit of Measure
	Labor ($K_L = 13.48$)	Rental ($K_{Cr} = 75$)		
18	$ 30	$ 15	—	Each
24	50	25	0.2	Each
30	65	30	—	Each
36	95	45	0.3	Each
42	140	65	—	Each
48	190	90	0.4	Each
54	240	120	—	Each

inches thick by 12 inches deep is poured at the end of the flared section. The unit costs shown in Table 6.13 include the costs of fine grading, of unloading the end sections, of incidental excavation and backfill, and to form and pour the typical cutoff walls.

6.4.8 Cutting Reinforced Concrete Pipe

Concrete pipe is ordered to exact lengths, and field cutting is avoided when possible (see Table 6.14 for some estimated costs).

6.4.9 Shaped Bedding

Shaped bedding is sometimes specified for pipes installed under roadways. When required, shaped bedding is normally bid as a separate job item at a price per square foot of shaped surface.

The shaping is accomplished with a template conforming to the outside diameter of the pipe. The template is guided by two 2 × 4 headers which are set to the required line and grade.

TABLE 6.14 Estimated Costs for Field Cutting Reinforced Concrete Pipe[a]

12	$ 12	0.5
24	20	1
36	40	2
48	80	3
60	190	4
72	300	8

[a]The rental costs for the following equipment are additional:

1. Electric hand saw with carborundum blade.
2. Air Compressor.
3. Paving breaker.

Indexed to K_{6+2} = 1000, the headers will be set at a cost of about \$0.75 per linear foot of header, and the template will cost about \$7 per foot of width. Indexed to K_L = 13.48 and K_{BH} = 25, the unit costs for shaping and compacting will be about \$0.32 per square foot for labor and about \$0.15 per square foot for equipment rental. In most cases, an allowance for template or header materials is not necessary.

6.5 MISCELLANEOUS COSTS

The unit costs for concrete removal and for a number of miscellaneous minor items and activities are covered in this section. The unit costs for the minor items are both interesting in themselves and useful for estimating unit costs for other work by comparison.

6.5.1 Concrete Removal

Subcontractors are always used when removal of large quantities of concrete is required. Skilled equipment operators, expert planning and job supervision, and past experience on similar demolition work are essential. Concrete removal is inherently risky, and a high degree of precision is not possible when estimating its cost. Because of this and because of differing experience backgrounds, the prices quoted by experienced demolition contractors, for a given job, will sometimes vary by as much as 100 percent.

On unit price contracts, concrete removal is normally bid as a separate job item at a price per cubic yard or as a lump sum. The price bid may include the costs of certain excavation and backfill and normally includes the costs to load and dispose of the broken concrete; the costs for such work are covered in Chapters 4 and 5.

Production rates and unit costs for concrete removal on several jobs are given here. These rates relate only to the costs for breaking concrete ready for loading with a $2\frac{1}{4}$- to 3-CY loader. Unit costs or production rates should never be blindly used, and this is particularly true where demolition work is involved.

6.5.2 Paving Breaker Work

When paving breakers are used for breaking concrete, the moil points must be replaced with sharp ones at about 1-hour intervals. The sharpening must be done by experienced blacksmiths since improperly annealed moil points break immediately.

Indexed to K_L = 13.48, assuming that the work surface is approximately horizontal and that an 80-pound paving breaker is used, reinforced concrete will be broken at a unit labor cost ranging from about \$84 to \$111 per CY. Assuming one and one-half laborers per paving breaker, the corresponding production rates will range from about 0.24 to 0.18 CY per hour.

If the paving breaker must be operated in a horizontal position, the preceding unit costs will increase from two to three times. In all cases, the rental costs for the compressor, hoses, and air tools, the cost for sharpening moil points, and the costs for staging or for supports for paving breakers working horizontally are additional.

6.5.3 Wrecking Ball Work

Skilled workmen demolished two reinforced concrete T-girder bridges using a crane equipped with a 5000-pound wrecking ball. Six hundred and ninety cubic yards of deck plus girders were broken at a rate of 8.3 CY per hour. The hourly cost of the five-man crew was:

1 labor foreman	13.48 + 1.35 = $14.83
1 laborer with cutting torch	13.48 + 0.25 = $13.73
2 laborers with paving breakers	2 × 13.73 = $27.46
1 laborer helping	$13.48
	$69.50

The unit cost for labor (indexed to $K_L = 13.48$) was

$$\frac{69.50}{8.3} = \$8.40 \text{ per CY}$$

Indexed to $K_{Cr} = 75$ and assuming the daily rental rate for the compressor and tools to be $50, the unit costs for the equipment were:

$$\text{Crane} \qquad \frac{75}{8.3} = \$9.00 \text{ per CY}$$

$$\text{Compressor} \quad \frac{6.25}{8.3} = \$0.75 \text{ per CY}$$

Seven hundred and eighty cubic yards of bridge columns and footings were broken at a rate of 9.0 CY per hour. The columns were approximately 3.5 feet square by 30 feet in height. A three-man crew was used; no paving breakers were required, although the compressor and tools were standing by.

Indexed to $K_L = 13.48$, $K_{Cr} = 75$, and a daily rental rate of $50 for the standby compressor and tools, the unit costs were:

$$\text{Labor} \qquad \frac{42.04}{9.0} = \$4.70 \text{ per CY}$$

$$\text{Crane} \qquad \frac{75}{9.0} = \$8.30 \text{ per CY}$$

$$\text{Compressor} \quad \frac{6.25}{9.0} = \$0.70 \text{ per CY}$$

Indexed to the same labor and rental rates, a skilled five-man crew broke 380 CY of 16-inch-thick flat slab bridge deck plus 95 CY of abutments and wingwalls at unit costs of:

Labor	$8.50 per CY
Crane	$8.80 per CY
Compressor	$0.75 per CY

Seventy reinforced concrete pile extensions 16 inches in diameter by 7 feet long were broken by the same crew at unit costs of:

Labor	$145 per CY, or $18 each
Crane	$126 per CY, or $15 each
Compressor	$ 13 per CY, or $1.60 each

Forty-seven cubic yards of 6-inch reinforced concrete deck slab plus 35 CY of abutments and wingwalls were broken with a crane equipped with a 5000-pound wrecking ball at a rate of 5.8 CY per hour. The crane operator was average, and one laborer was used as a burner. No paving breakers were required. Indexed to $K_L = 13.48$ and $K_{Cr} = 75$, the unit costs were:

$$\text{Labor} \quad \frac{13.48 + 0.25}{5.8} = \$\ 2.40 \text{ per CY}$$

$$\text{Crane} \quad \frac{75}{5.8} = \$13.00 \text{ per CY}$$

Indexed to $K_L = 13.48$, $K_{Cr} = 75$, and a daily rental rate for a compressor and paving breaker of $50, forty-four reinforced concrete headwalls averaging 1.38 CY each were cut loose and broken for unit costs of:

Labor	$37 each, or $27 per CY
Crane	$56 each, or $41 per CY
Compressor	$11 each, or $ 8 per CY

6.5.4 Hydraulic-Impulse Breakers

These units weigh about 1000 pounds and are mounted on the boom of a small rubber-tired loader-backhoe. They operate from the tractor's hydraulic system, and no compressor is required.

Forty-two cubic yards of reinforced concrete consisting of the outer 3 feet of a 12-inch-thick bridge deck overhang and a 2-foot walkway 8 inches thick on a deck slab which was to remain were removed. One and one-half laborers, one paving breaker, and one hoe-mounted impulse breaker were used.

Indexed to $K_L = 13.48$, an hourly rental rate for the impulse breaker of $40, and a daily rental rate for the compressor and paving breaker of $50, the unit costs were:

Labor	$51 per CY
Impulse breaker	$67 per CY
Compressor	$ 7 per CY

In all cases, when portions of a reinforced concrete structure are to be removed, extra care and time are required to avoid damaging the remaining structure.

6.5.5 Concrete Sawing

Saw cutting is frequently required when portions of a concrete structure must be removed. This work is usually done by subcontractors using self-propelled concrete cutting saws equipped with diamond blades, and the same equipment is used to saw cut weakened-plane joints in new concrete pavement.

Saw cuts in old unreinforced concrete slabs are made at about the following speeds:

Depth of Cut	Linear Feet per Hour
$1\frac{1}{2}$	55
2	50
3	40
4	25

Actual saw-cutting speeds vary with the quality of the concrete and with the amount of reinforcing steel encountered. With No. 4 or 5 bars at 12-inch centers, the cutting speeds will range from about 35 to 50 percent of those shown in the preceding. When asphalt concrete or fresh portland cement concrete is cut, the speeds will be about 6 times those shown.

Vertical and horizontal cuts in reinforced concrete walls are usually short and require more time. The self-propelled saw used is lighter in weight and operates from a temporary rail which must be bolted to the wall. Door openings 3 × 7 feet will be cut in 4-inch reinforced concrete walls in about 4 hours, and about 6 hours are required for such openings in walls 7 inches thick.

6.5.6 Pavement Removal

A 185-HP tractor ripped $8\frac{1}{2}$-inch-thick asphalt concrete pavement at an effective rate of about 550 square yards per hour. Cracked portland cement concrete pavement 8 inches thick was ripped at an effective rate of about 340 square yards per hour. In both cases, the material was loaded with a track-type loader having a $2\frac{1}{2}$-CY bucket.

Portland cement concrete pavement 8 inches thick was broken by a crane equipped with a 5000-pound wrecking ball at an effective rate of about 139 square yards per hour. Using a self-propelled, 1500-pound hydraulic hammer, the same pavement was broken at an effective rate of about 220 square yards per hour. In both cases the material was loaded with a track-type loader having a $2\frac{1}{2}$-CY bucket.

A crane equipped with a wrecking ball broke concrete curb, gutter, and sidewalk at an effective rate of about 77 CY per hour.

6.5.7 Minor Items and Costs

The costs for the following items are, in most cases, small in comparison with the total job cost. The data presented represents average costs when incidental amounts of such work are done by the general contractor's own forces.

6.5.8 Markers and Reflectors

A typical highway marker consists of a 10-gauge galvanized post 3 inches wide by 6 feet long having a single vertical corrugation and weighing about 15 pounds. The posts are driven to a penetration of 2 feet at specified locations. An enameled plate 8×24 inches weighing about 7 pounds is attached to the post with eight $\frac{1}{4}$-inch bolts.

Indexed to $K_L = 13.48$, the labor cost to install markers, complete, is about $4.65, and no equipment rental is involved.

6.5.9 Right-of-Way Monuments

A typical right-of-way monument is a portland cement concrete post 6 inches square by 3 feet 6 inches long weighing about 125 pounds. The posts are set 3 feet deep at specified locations.

Indexed to $K_L = 13.48$, the labor cost to set each prefabricated monument is about $19.

6.5.10 Survey Monuments

Survey monuments are of two types. A type 1 monument is a concrete post about 15 inches square by 24 inches long, weighs about 450 pounds, and is set flush with the ground. A type 2 monument is about 15 inches square by 18 inches long, weighs about 370 pounds, and is set 28 inches deep. A short piece of 6-inch-diameter pipe with a cover is set on top of the post.

Indexed to $K_L = 13.48$, the labor cost to set either prefabricated monument will be about $20. Type 1 monuments, cast in place, will cost about $30 each for labor.

6.5.11 Metal Beam Guard Rail

A metal beam guard rail consists of a corrugated metal rail bolted to wood posts. The galvanized rail panels are about 12 inches wide by 13 feet 6 inches long, have two $3\frac{1}{4}$-inch corrugations, and weigh about 108 pounds. The treated posts are 8 × 8 rough by 5 feet 4 inches long and are embedded 3 feet deep at 6-foot-3-inch or 12-foot-6-inch centers. An 8 × 8 filler block 14 inches long is used at each post, and the rail is attached with one $\frac{5}{8}$-inch bolt per post. The rail panels are lap spliced at the posts, and eight $\frac{5}{8}$-inch bolts are used for each splice.

Indexed to $K_{2+3} = 585$, where K_{2+3} = the total cost for two carpenters plus three laborers working for 8 hours, the unit labor costs for installing a metal beam guard railing are about as follows:

6-foot-3-inch post spacing: $4.16 per linear foot.
(140 linear feet per day)
12-foot-6-inch post spacing: $3.92 per linear foot
(150 linear feet per day)

In both cases, a hand-operated, power-driven post hole auger is required; when large quantities are involved, a truck mounted auger may be used.

6.5.12 Right-of-Way Fence

Fencing is always installed by subcontractors when the quantities are large. Two laborers installed five-strand barbed wire fence on corrugated metal posts spaced 12 feet on center at a rate of 480 linear feet per day. Indexed to $K_L = 13.48$, the unit labor cost was $0.45 per linear foot. Using wood posts, a similar fence was installed for a unit labor cost of $0.95 per linear foot, and a hand-operated, power-driven post hole auger was required.

6.5.13 Raised Traffic Bars

Raised traffic bars are prefabricated of portland cement concrete, are 8 inches wide by $1\frac{1}{2}$ to 2 inches high, and weigh 9 to 14 pounds per linear foot. they are installed on a thick trowel coat of asphalt adhesive and are given one coat of traffic paint on completion.

Indexed to $K_L = 13.48$, small quantities of raised traffic bars were installed, complete, at a unit labor cost of $1.10 per linear foot.

6.5.14 Barricades

Indexed to $K_{6+2} = 1000$, portable timber barricades 4 feet wide by 5 feet high were made from 2-inch material at a unit labor cost of $260 per MFBM. Barricades 12 feet wide by 5 feet high were made from 2- to 4-inch material at a unit labor cost of $270 per MFBM. In both cases, the costs include the application of one coat of paint.

On unit price contracts, the price bid for barricades normally includes the costs for furnishing, placing, repairing, or replacing damaged units. In general, the longer the barricades are in use, the more damage they sustain, and on typical freeway work, more than two barricades must be furnished, over the life of the job for each one covered by the job item.

6.5.15 Roadside Signs (Wood Posts)

The sign panels for roadside signs are normally furnished, FOB job, by the owner. The price bid for roadside signs includes the furnishing of all wood posts, braces, and hardware and the costs of installing the posts and sign panels, complete, in place. The unit of measure for the item is thousands of board feet.

Indexed to $K_{16+2} = 1000$, typical roadside signs having 4×4 to 6×8 wood posts were installed at an average unit cost for labor of $1070 per MFBM, not including the cost of painting. A hand-operated, power-driven post hole auger was used. Indexed to $K_L = 13.48$, the unit labor cost for painting the posts was about $330 per MFBM for three coats.

6.5.16 Roadside Signs (Steel Posts)

This item is similar to the wood posts except that the unit of measure is pounds of steel in the posts. In addition, the steel posts are normally backfilled with 4 sack concrete.

Indexed to $K_{6+2} = 1000$ and $K_{Cr} = 50$, roadside signs on steel posts were installed at the following unit costs:

> Labor $96 per post, or $0.16 per pound
> Crane $33 per post, or $0.055 per pound

A truck-mounted post hole auger was used; it averaged four holes per hour.

6.5.17 Remove Roadside Signs (Wood Posts)

Indexed to $K_L = 13.48$, roadside signs with wood posts were removed and stockpiled at a unit cost for labor of $8.00 per post. No equipment was required.

6.5.18 Sacked Concrete Slope Protection

Sacks made of 10-ounce burlap measuring about $19\frac{1}{2}$ inches wide by 36 inches long are filed with 1 cubic foot of 4 sack concrete and placed on the slope as specified.

This is heavy work, and it is imperative that the concrete be delivered close to the location where the sacked concrete will be placed.

Using a portable hopper equipped with a job-made, 1-cubic-foot metering box, ten laborers placed several hundred cubic yards of sacked concrete at an average rate of 3.1 CY per hour. Two three-man crews were used to carry and place the sacks, and the average carry distance was about 15 feet down a 2:1 slope. One laborer was used grading and filling at the point of depositing, and three laborers filled the sacks at the hopper and handed them to the carrying crews.

Indexed to K_L = 13.48, the unit labor cost for this work was $44 per CY.

When small quantities are involved and the concrete can be chuted directly to the point of deposit from the truck, a four-man crew will do the work at a slower rate for the same unit cost. These unit costs do not include the costs for excavation, which are normally paid for under a separate job item.

6.5.19 Sand Bags

Three laborers can fill and place about 40 bags per hour if the bags can be deposited close to the point of filling. The bags contain about 0.9 cubic feet of sand and will cover about 1.8 SF per bag on the level and about 1.4 SF per bag on a $1\frac{1}{2}$:1 slope.

6.5.20 Grouting Broken Concrete Riprap

Indexed to K_L = 13.48, broken concrete riprap was grouted with concrete which was chuted to the required spot at a unit labor cost of $12 per CY. This unit cost includes the costs of spading the grout into the spaces between the pieces of broken concrete, finishing the surface so the faces of the broken pieces are exposed, and curing the grout.

6.5.21 Rock Slope Protection

Indexed to K_L = 13.48 and K_{C_r} = 75,270 CY of $\frac{1}{4}$-ton rock slope protection was placed at unit costs of:

Labor $1.40 per CY
Rental $4.20 per CY

The material was end dumped over a bank and placed with a $\frac{3}{4}$-CY clamshell bucket.

Indexed to the same rates, 270 CY of excavation for the rock was done at unit costs of:

Labor $0.65 per CY
Rental $1.60 per CY

Four hundred tons of rock backing consisting of rocks weighing from 25 to 50 pounds each were placed with a 1-CY clamshell bucket at an effective rate of 40 tons per hour. Seven hundred tons of 1-ton rock slope protection was placed over the backing with the same equipment at an effective rate of 43 tons per hour. In both cases, the material was placed on a relatively inaccessible slope behind a concrete pile bent which was about 15 feet in height. Working on banks approximately 8 feet in height, both the backing and the 1-ton rock were placed with a $2\frac{1}{4}$-CY wheel loader at an effective rate of 60 tons per hour.

6.5.22 Dry Pack Window Mullions

Window mullions requiring approximately 2 cubic inches of dry pack per mullion were dry packed at a unit cost of $2.86 each indexed to $K_F = 16.10$. This is equivalent to an effective rate of

$$\frac{16.10}{2.86} = 5.6 \text{ mullions per hour}$$

and this rate is typical for many repetitive operations of this type, where the workman must go to a location, do a minimum amount of work, clean up, and move to the next nearby location.

6.5.23 Welding

Pieces of iron pipe 2 inches in diameter by 10 inches long were welded to the top flanges of plate girders for use as screed supports. The deck forms were in place, and access was easy. The pipes were set 4 feet on center, and four spot welds were made for each pipe. A good welder attached the pipes at a rate of 36 pipes per hour.

Angular iron clips were welded to the top flanges of the outside girders at 4-foot centers. Eight inches of $\frac{1}{4}$-inch fillet weld was used for each bracket, and a good welder did the work at a rate of 7 brackets per hour.

6.5.24 Remove Wooden Rail Posts

Eight-by-eight bridge rail posts 5 feet long were removed and stockpiled for salvage. Each post had four rusted bolts which were burned and driven out with drift pins. Two laborers did the work at a rate of 2.8 posts per hour.

6.5.25 Wood Box Cofferdam

One hundred fifty linear feet of wood box cofferdam filled with earth was used to dry up one side of an unlined canal. The water was 4 feet deep.

The box section was 4 feet wide by 5 feet high and contained approximately 110 CY of dirt. Two-by-twelve planks were used for the box sides. Four-by-six vertical wales together with two 4 × 4 struts were used at 4-foot centers, and the wales were tied together with twisted No. 9 wire. The open-ended box sections averaged 14 feet in length, and the entire cofferdam required 4.5 MFBM of material, which is equivalent to 3 board feet per square foot of box wall.

Indexed to $K_{6+2} = 1000$ and $K_{Cr} = 75$, the unit costs were:

Make box sections:

Labor $250 per MFBM

Set box sections and fill with dirt:

Labor $430 per MFBM
Rental $230 per MFBM

Remove box sections and dirt:

Labor $310 per MFBM
Rental $340 per MFBM

The dirt fill was placed and partially removed with a $\frac{3}{4}$-CY clamshell bucket. The final cleanup was made with a $\frac{3}{4}$-CY dragline bucket.

6.5.26   Drilling Dowel Holes in Concrete

The cost to drill holes in existing concrete structures depends on the amount of reinforcing steel encountered. Using a light jackhammer, holes $1\frac{1}{2}$ inches in diameter were drilled vertically in unreinforced concrete at a rate of 120 inches per hour. Similar holes were drilled horizontally in lightly reinforced concrete at a rate of 48 inches per hour and in heavily reinforced concrete at a rate of 26 inches per hour.

One laborer will dry pack dowels in 24-inch-deep drilled holes at a rate of about 6 dowels per hour.

Note: Labor and material costs as well as unit costs used are for illustration only and do not reflect present-day costs.

CHAPTER 7

Piles

Pile driving requires highly specialized equipment, crews, and experience. On all building work and on most engineering work driven piles are furnished and installed by pile-driving contractors on a subcontract basis. In our estimates we assume all driven piles to be furnished and installed by subcontractors.

Many pile-driving contractors perform as general contractors on jobs where pile driving, heavy timber work, or underwater work constitute a major portion of the work to be done. This is commonly the case on waterfront work and on bridges where wide bodies of water are involved.

The general contractor will sometimes drive wood piles with his own forces when a few piles are required for temporary structures at a remote location. In such cases, the work is done with a truck crane equipped with a set of hanging leads and a drop hammer. A three-man inexperienced crew will drive 30-foot piles to a penetration of 20 feet in sandy loam soil at a rate of 8 to 10 piles per 8-hour shift.

Indexed to $K_{6+2} = 1000$, the piles will be cut off to grade at a cost of about \$7 per cut. In addition, about 6 hours will be required to rig up and dismantle the leads and hammer.

7.1 PILE EXTENSIONS

When concrete pile bents are required for structures built on dry land, the top of the driven portion of the piles is often set at 1 foot below the finished ground surface. The exposed portions of the piles are called pile extensions and are formed and poured as columns.

The driven portion in such cases usually consists of a steel shell or pipe filled with concrete. The combination of such piles and their extensions costs less than full-length precast concrete piles. In addition, the use of pile extensions enhances the appearance of the exposed bent since perfect alignment of the extensions can usually be achieved.

Pile extensions are usually constructed by the general contractor, and unit costs for this work is presented in Table 7.1

7.2 CAST-IN-DRILLED-HOLE CONCRETE PILES

Cast-in-Drilled-Hole (CIDH) concrete piles are frequently used in bridge construction when soil conditions are favorable and when the water table is known to be below the tip elevation of the piles. When holes must be drilled below the water table, casing and pumps must be used, and the cost advantage of CIDH piles over driven piles is lost. In rare cases, the underground water may be under pressure. The drilled holes then act as artesian wells, and the installation of CIDH piles in such circumstances is virtually impossible.

When large amounts of CIDH piles are required, the work is normally done by drilling subcontractors using crane-mounted auger-type drills. Where small quantities are involved at locations far from drilling contractors, the drilling can be done with truck-mounted rotary bucket drill rigs, which are commonly available.

Unit costs for CIDH pile work based on using a bucket drill rig and dry holes are given in Table 7.1, and their use is illustrated in the cost estimate for the flat slab bridge. For full information see *Construction of Drilled Pier Foundations* by D. M. Green, P.E., and W. S. Gardner, P.E., a volume in this series.

TABLE 7.1 Unit Costs for Cast-in-Drilled-Hole Piles and Pile Extensions

	Labor		Rental		
Activity	Unit Cost ($)	Index	Unit Cost ($)	Index	Unit of Measure
16″ φ CIDH piles					
Pour directly from truck	6.00	$K_L = 13.48$	—	—	CY
Drill, set reinforced steel, and pour	0.60	$K_L = 13.48$	1.46	$K_{BD} = 39$[a]	LF
16″ φ Pile extensions					
SS sonotube, set reinforced, steel, and pour	5.65	$K_{6+2} = 1000$	—	—	LF
Or:					
SS sonotube	3.80	$K_{6+2} = 1000$	—	—	LF
Set reinforced steel cages	0.29	$K_L = 13.48$	—	—	LF[b]
Pour from deck forms[c]	1.55	$K_L = 13.48$	—	—	LF
Pour with crane	12.00	$K_L = 13.48$	28.00	$K_{Cr} = 75$	CY

[a]K_{BD} = hourly rental rate for truck-mounted, rotary bucket drill rig, operated and maintained. On-and-off charges are additional.

[b]Per linear foot of cage.

[c]Includes the costs to buggy concrete for bridges about 100 feet in length.

Note: Drilling and pouring costs based on linear feet are directly proportional to the square of the hole diameter. When 11 CY or more of pile extension are poured at one time, the unit cost for the crane will include 1 hour of travel time.

7.3 CIDH PILES

The estimate sheet for this work is shown in Figure 7.1, and the unit costs for doing this work are shown in Table 7.1.

The pay length for CIDH piles is measured from the top of the pile extension to the bottom or tip of the pile. Two types of work are involved, and the quantities are taken off and estimated separately as shown. This allows a quick comparison to be made on bid day if a drilling subcontractor submits an unforeseen bid for CIDH portion of the work.

The reinforcing steel for CIDH piles is not included in the job item for reinforcing steel, and its cost must be included in the cost of the pile item. The reinforcing steel for pile extensions normally consists of six No. 6 vertical bars wrapped with W3.5 wire with a 6-inch pitch. The vertical bars extend 18 inches into the concrete above. The reinforcing steel for the CIDH portion normally consists of four No.6 bars wrapped as before, with the vertical bars extending 18 inches into the pile

ITEM 5. 683 LF CIDH PILES (682.5)				
L 13.48 K$_{6+2}$ 1000 C 16.34 F 16.10 BUCKET DRILL 38		L	DAYS	M, R

S$_1$ 525 LF CIDH
CONC. 27.1 × 1.08 = 29.3 CY   30
REINF. CAGES 283.5 LF   2
DRILL, SET CAGES & POUR
   525 LF   0.60   1.46
B. DRILL ON-OFF 2 HRS.   38
 	
        INCL. PAY ROLL EXP. (1.305)
  Σ$_1$ = $2700   ($5.14/LF)
 	
S$_2$ 157.5 LF PILE EXTENSION
CONC. 8.1 × 1.02 = 8.3 CY   30
REINF. CAGES 189 LF   2.80
168 LF 16"ϕ SONOTUBE   3.75
SS SONO., SET CAGES
   AND POUR 157.5 LF   5.65
 	
        INCL. PAY ROLL EXP. (1.315)
  Σ$_2$ = $2580   ($16.38/LF)
 	
  Σ$_1^2$ = $5280
		315
		—
		315
		410
		
		890
		890
		1170
			765
			8×38
			= 3D
			
			890
			1000×⅓
			= 3D
			
			Σ= 6D
				880
				565
				—
				765
				80
				2290
				
				250
				530
				630
				
				1410

Figure 7.1. CIDH pile costs.

extension. The reinforcing steel for the CIDH portion may or may not extend to the full depth of the pile. In our case, we assume these cages are specified to extend 12 feet below the joint, and the total length of these cages will be 13.5 feet each.

Since the linear feet of required cages is never equal to the linear feet of piles plus extensions, it is important for the contractor to determine exactly what the steel subcontractor is quoting when the cages are bid on a per foot basis.

The cost of disposing of the material excavated from the drilled holes is not included in the unit costs shown in Table 7.1. In our case, we assume this material may be left in the channel bottom.

The time estimate for the item is made in the days column of the estimate sheet. The time to drill and pour the piles is determined by dividing the rental cost of the drill rig by the daily rental rate, and amounts to 3 days. The 3 days required to do the pile extensions are based on the total labor cost and an assumed two-carpenter crew.

Note: Labor and material costs as well as unit costs used are for illustrations only and do not reflect present-day costs.

Structure Concrete

8.1 ENGINEERING WORK (HEAVY CONSTRUCTION)

The quantities of all materials, labor, and equipment that apply to the structure concrete item as a whole, together with the quantities of certain other minor activities, are collected and listed at the beginning of the estimate sheet for the structure concrete item in a section designated "General Costs." The costs of some of these items depend upon the time required to complete the structure concrete work, but none of them are normally on the critical path for the structure concrete item.

Separating the general costs from those labor and equipment costs directly related to performance of the work simplifies the estimate and the monitoring of costs on work in progress, decreases the possibility of error, and facilitates determining the time required for completion.

The separate operations required to complete the structure concrete item together with the quantities of their related primary activities are then listed below the general costs section in a logical sequence like that illustrated in the estimates that follow. The activities so listed are priced and the time required to complete the item is determined.

With the time required to complete the structure concrete item established, the general costs section is priced and the total estimated cost of the structure concrete item is obtained.

On some jobs there may be more than one job item for structure concrete. In such cases, a single combined takeoff is made for all the structure concrete work and a single cost is determined for the total quantity of concrete involved. This procedure is necessary because many of the general costs relate to the total quantity of structure concrete on the project and cannot be realistically assigned to a particularly bid item. When the total cost of the combined structure concrete items has been determined, the unit prices bid for the individual items are assigned somewhat arbitrarily, as illustrated in the estimates that follow.

8.1.1 General Costs Checklist

Those items from the following list that are applicable to the given job are listed and priced in the general costs section of the structure concrete estimate. Such listing and pricing requires very little time when a good takeoff of the structure concrete item has been made.

1. Concrete.
2. Form lumber.
3. Falsework lumber.
4. Steel shoring.
5. Steel beams.
6. Column forms.
7. Concrete hoist.
8. Scaffolds and staging.
9. Form ties.
10. Minor expenses.
11. Make forms.
12. Additional finishing.
13. Additional curing.
14. Sandblasting.
15. Set anchor bolts.
16. Dry pack anchor bolts.
17. Compressor.
18. Vibrators.
19. Table saw.
20. Finishing machine.
21. Develop power.
22. Develop water.
23. Hoppers and buggies.
24. Neoprene-bearing pads and expansion joint material.
25. Curing paper.
26. Curing blankets.
27. Miscellaneous iron.
28. Deck and screed hardware.
29. Box girder dobies.
30. Clean girders.

31. Sweep bridge decks.
32. Clean up and out.
33. On-and-off charges for steel falsework.
34. Other.

8.1.2 Concrete

This item covers the purchase price, FOB job, of the portland cement concrete required. In the estimates that follow it is assumed that the concrete is available in transit-mix trucks.

An allowance of 2 percent for waste is adequate for all formed concrete and for most jobs taken as a whole. Where grade slabs constitute a large portion of the work, an allowance of 5 percent for waste should be used for the grade slabs. Cast-in-drilled-hole piles require an allowance of 7 to 10 percent for waste.

8.1.3 Form Lumber

This item covers the purchase price, FOB job, of the form lumber and plyform required. The amounts required depend on the number of reuses that are possible. Estimation of form lumber requirements is based on the concrete takeoff and is illustrated in the estimates that follow.

Wall form panels are assumed to be made with 2×4 studs and double 2×4 wales with form ties at 2 feet on center both ways. Plyform is assumed to be $\frac{5}{8}$ inches thick. Studs are assumed to be 12 inches center to center, except that for walls or girders under 5 feet high, they may be on 16-inch centers.

In our estimates we allow 1.75 to 2 MFBM of 2×4 lumber and 1.17 MSF of plyform for each MSF of wall form panels required.

When contractor-owned used material is available, it will be charged to the job at one-third the cost of new material, and this charge will include the cost of moving the material to the job.

8.1.4 Falsework Lumber

This item covers the purchase price, FOB job, of the falsework lumber required. When contractor-owned used falsework lumber is available, it will also be charged to the job at one-third the cost of new material, and this charge will include moving it to the job.

The amount of material required is based on a preliminary design for the falsework in all cases. Where large structural slabs with various arrangements of beams and girders occur, the falsework required for a typical bay is determined, and the total falsework required is estimated from the ratio of the total area to be formed to the area of the typical bay.

8.1.4.1 FALSEWORK DESIGN. In our estimates we use the following loads and working stresses when designing falsework.

The minimum total design load will not be less than 100 pounds per square foot regardless of slab thickness.

Subject to the preceding minimum, the design load will be the greater of the loads determined by the following two methods:

1. One hundred sixty pounds per cubic foot for concrete plus the actual weight of any equipment supported plus 20 pounds per square foot plus 75 pounds per linear foot applied at the outside edge of deck overhangs.
2. One hundred fifty pounds per cubic foot for concrete plus the actual weight of any equipment supported plus 50 pounds per square foot plus 75 pounds per linear foot applied at the outside edge of deck overhangs.

Horizontal loading will be the greater of the wind load, or 2 percent of the total dead load.

Deflection will be limited to $\frac{1}{240}$ of the span.

The maximum stresses will be as follows:

Timber	
Compression perpendicular to the grain	450 psi
Compression parallel to grain but not to exceed 1600 psi	$\dfrac{480,000}{(l/d)^2}$ psi
l = unsupported length of column in inches	
d = least dimension of column in inches	
Flexural stress	1500 psi
Horizontal shear	140 psi

In calculating horizontal shear stresses, all loads within a distance from either support equal to the depth of the beam will be neglected.

Nailed joints are common in constructing formwork. Whenever possible, scaffold nails are used since they have a double head and are easily removed during the stripping operation. Nail sizes are designated by pennyweight, which is abbreviated as d. For a given pennyweight, box nails are smaller in diameter than common wire nails, furnish less lateral resistance, and are not normally used in construction since they bend easily when driven. Nails may be bright, galvanized, or cement coated.

The properties of wire nails are given in Table 8.1. In our designs we use 8d nails for fastening surfaced 1-inch material and 16d nails for surfaced 2-inch material. Assuming scaffold nails driven in the side grain of coastal region douglas fir, we use the following lateral resistances:

8d scaffold nail	70 pounds each
16d scaffold nail	100 pounds each

TABLE 8.1 Properties of Wire Nails

Size (d)	Length (inches)	Wire Gage Number	Approximate Number per Pound
		Bright Common Nails	
4	$1\frac{1}{2}$	$12\frac{1}{2}$	314
6	2	$11\frac{1}{2}$	181
8	$2\frac{1}{2}$	$10\frac{1}{4}$	106
16	$3\frac{1}{2}$	8	49
		Bright Scaffold Nails[a]	
6	$1\frac{3}{4}$	$11\frac{1}{2}$	150
8	$2\frac{1}{4}$	$10\frac{1}{4}$	88
16	3	8	44

[a]The lengths shown for scaffold nails are the net length below the lower head.

Steel: All structural steel falsework will be designed in accordance with the Manual of Steel Construction published by the AISC.

8.1.5 Steel Shoring

This item covers the rental cost of heavy-duty steel shoring similar to that shown in Figure 8.1. Such shoring is frequently used for bridge falsework. Under normal conditions, the cost to set and strip steel shoring is less than the cost to set and strip wood falsework bents. In addition, the danger of a falsework failure due to undetected defects in rough wood posts is eliminated.

The load-carrying capacity and the weight of the shoring towers must be obtained from the manufacturer in all cases. In our designs we assume an allowable load of 10 kips per leg, and our setting and stripping costs will be based on a unit price per CWT for shoring of a given height. We assume such shoring is contractor owned and charge the jobs a monthly rental rate of $2 per CWT of shoring for those months the shoring is in use.

8.1.6 Steel Beams

This item covers the rental charged for steel wide-flange beams used for falsework. In our estimates we assume the beams are owned by the contractor, and we charge a rental of $1.50 per CWT per use.

8.1.7 Column Forms

This item covers the purchase price, FOB job, of sonotube and the rental charged for steel column forms.

Figure 8.1. Shore "X" Tower. The trademark Shore "X" is registered by Waco Scaffold & Shoring Co.

8.1.8 Concrete Hoist

This item covers the monthly rental charges for concrete hoists or conveyors and includes the costs of moving the equipment on and off the job and of operating and maintaining the equipment. Concrete hoists are used primarily on building work and are discussed in Section 8.2.5. The hourly rental costs for truck-mounted concrete pumps or truck cranes used to place concrete are carried in the operating section of the estimate sheet, as illustrated in the following estimates.

8.1.9 Scaffolds and Staging

This item covers the monthly rental charges for portable scaffolds, scaffold jacks, and lightweight metal staging as well as the material costs for lumber when wooden staging is used. The cost to erect and dismantle the staging is carried in the section of the estimate requiring the staging.

8.1.10 Form Ties

This item covers the purchase price, FOB job, and/or the rental charged for wall form ties. Our estimates are based on using contractor-owned taper ties at 2 feet on center both ways. We charge the jobs a monthly rental rate for those months that the ties are being used. The monthly rental rate will be equal to or less than one-twelfth the cost of a new tie, and the maximum rental charged on any job will never exceed the purchase price of new ties. Typical form ties are shown later in Figure 8.36.

8.1.11 Minor Expenses

This item includes the costs of a number of items common to all concrete work. Included are the costs of nails, hand tools and incidental supplies, electric hand tools, saw filing, curing compound, form oil, and millwork. Individually, none of these items is of major importance, but as a whole they are significant, and an allowance must be made on the estimate for their cost. It is doubtful that any two contractors handle these items in exactly the same manner.

These items represent the costs for materials and services, and each of them may vary considerably from job to job. They do not relate directly to wage rate indexes, nor to any of the general economic indexes available. In practice, where recent job cost records are available, the group as a whole may be bid with sufficient accuracy at a unit cost per MCY of concrete involved. The unit costs developed will vary with the type of work being done.

Since the student will not have access to a file of recent job costs, we discuss each item individually and derive a unit cost for the group based upon assumed current prices and the conditions involved on the jobs we bid in the following sections.

Nails. This item covers the cost, FOB job, of the nails required to make, set, and strip the forms. The quantity of nails required per MCY of concrete varies with the number of reuses of the form panels, the complexity of the job, the relative amounts of wall, slab, and foundation concrete, and the thickness or mass of the structural sections. In our estimates we allow 10.5 CWT of nails per MCY of concrete on engineering work and 24 CWT of nails per MCY of concrete on concrete building work.

Hand Tools and Incidental Supplies. This item covers the costs of hand tools, such as shovels and pry-bars, together with minor supplies such as curb stakes, rubber boots, corks, and No. 9 wire. In practice, many of these items are obtained from the contractor's yard, and the job is only charged for the new ones purchased.

In our estimates we use a cost of $500 per MCY for this item, which relates roughly to a wholesale price index of 200 (year 1967 = 100).

Electric Hand Tools. This item covers the rental charged to the job for electric saws, drills, grinders, impact drills, and the like, that are required together with the electric cords required for their use. In our estimates we charge a monthly rental of one-tenth the cost of the tools involved and assume two saws and one-half drill will be required by each six-carpenter crew.

Saw Filing. This item covers the cost of filing the carpenter's hand saws as well as the electric saw blades. In our estimates, we assume each six-carpenter crew will require three handsaws and two electric saw blades to be sharpened for each week the crew works and that the cost to sharpen either one will be one-fifth of the hourly wage rate for a carpenter. It is assumed that the filing will be done by a saw-filing shop.

Curing Compound. This item covers the cost, FOB jobs, of the curing compound and the sprays required to apply it. The labor cost to apply the curing compound is included in the cost of pouring the concrete. In our estimates, we plan to use garden-type hand spray cans and allow 1 gallon of curing compound for each 150 square feet of surface to be sprayed.

Form Oil. This item covers the cost, FOB job, of the form oil and the sprays required to apply it. The labor cost to apply form oil and the labor cost to clean forms for reuse are included in the costs of making, setting, and stripping forms. In our estimates we plan to use hand spray cans as before and allow 1 gallon of form oil for each 150 square feet of formed surfaces.

Millwork. This item covers the cost, FOB job, of $\frac{3}{4}$ inch chamfer strip and of special millwork items such as wedges or form-marking strips when required. In our estimates we may take off the chamfer strip with the concrete and charge it at $0.13 per linear foot for the material. The labor cost to apply chamfer strip is included in the cost to set and strip forms. If marking strips are required on the form panels, a labor cost of $0.15 per linear foot of marking strip should be used based on $K_{6+2} = 1000$, and this labor cost should be included with the cost to make the form panels.

8.1.12 Make Forms

This item covers the cost to make the required wall form panels. The amount or area of forms to be made is determined on the concrete takeoff, as shown in Section

8.1.37.1. Form-marking strips, when required, represent an additional expense, as noted in Section 8.1.11.

8.1.13 Additional Finishing

This item covers the costs for the labor and materials required to fill form tie holes, remove fins, patch rock pockets, and grind or sack surfaces. The amount of additional finishing required will depend on the quality of formwork and on the job specifications.

The total costs for additional finishing are fairly uniform for a given type of structure and are generally proportional to the amount of concrete involved. In our estimates we assume first-class formwork is used, and we estimate the cost of additional finishing based on a unit cost per MCY of concrete in a particular type of structure in many cases. When necessary, the quantities involved will be taken off, and the unit costs presented in Table 8.2 will be used to determine the cost of the additional finishing.

When the soffit formwork for structural slabs and girders has been properly made, very little touch-up is required for the soffits. Therefore the areas of structural slab and girder soffits are not included in the square footage used to estimate the costs to touch up and patch.

The cost to fill form tie holes constitutes a substantial portion of the additional finishing cost. It is important that no costs be charged where form tie holes do not need to be filled, as may be the case in connection with some foundation walls and the interior girders of box girder bridges.

8.1.14 Additional Curing

This item covers the costs to water cure concrete on weekends and holidays. The normal cost for curing on working days is included in the cost to pour the concrete. In the rare case where the specifications require one or more men to be assigned to

TABLE 8.2 Unit Costs for Additional Finishing Work

Activity	Unit Cost for Labor	Index	
Fill form tie holes	$600–$650 per thousand holes each face[a]	K_{L+F}	= 29.58
	$300 per thousand holes each face[b]	K_L	= 13.48
Grind for uniform color	$270 per MSF	K_F	= 16.10
Sack nearby walls	$500–$575 per MSF	K_F	= 16.10
Sack beam sides and			
beam and slab soffits	$260 per MSF	K_F	= 16.10
Touch up and patch	$30 per MSF formed[c]	K_F	= 16.10

[a]Exposed, nearby wall surfaces. While cement is used, and patches are not visible on completion.
[b]Unexposed surfaces. Patches visible upon close inspection.
[c]Exclude soffit areas.

Note: For estimating purposes, assume the costs for materials equal 4 percent of the cost for labor.

curing on a full-time basis, the entire cost for such labor should be included under this item.

8.1.15 Sandblasting

This item covers the labor and material costs for sandblasting construction joints. In our estimates, we use a unit cost per MCY of concrete in some cases and a unit cost per square foot in others.

Where the areas to be sandblasted at any one time range from 50 to 800 SF, we use a unit cost for sandblasting labor of $0.30 per SF indexed to $K_L = 13.48$. The sand used must be dry. We assume it is purchased in 100-pound sacks and that 150 pounds of sand will be used per hour. The rental cost for the sandblast pot, hose, and nozzle will be included in the rental charged for the compressor.

The cost to water cut laitance from freshly poured construction joints is included in the unit costs to pour concrete.

8.1.16 Set Anchor Bolts

This item covers the labor cost to set anchor bolts for heavy building columns or for bridge girder connections. Such bolts may be $1\frac{1}{4}$ inches in diameter by 24 inches long, will usually have two or four bolts in each set, and will often be furnished with plate washers and pipe sleeves.

In our estimates we use a unit cost for labor of $24 per two-bolt set and $44 per four-bolt set based on $K_{6+2} = 1000$. These unit costs will include the costs to make, set, and strip the bolt templates and to set and adjust the bolts.

The material cost of the anchor bolts is included in the cost of the job items for structural steel or miscellaneous iron in all cases.

8.1.17 Dry Pack Anchor Bolts

This item covers the cost to dry pack the anchor bolt pipe sleeves and the column base plates or masonry plates. We use a unit cost of $50 per cubic foot to place dry packing based on $K_{L+F} = 29.58$. In addition, an allowance for materials will be made based on using 1 part of cement to $2\frac{1}{2}$ parts of concrete sand for the dry pack mix.

8.1.18 Compressor

This item covers the rental charges for a 105- to 120-CFM portable air compressor and includes the charges for small air tools, air hose, and the sandblast pot.

8.1.19 Vibrators

This item covers the rental charges for the concrete vibrators. In our estimates we plan to have a backup vibrator available for each pour. The labor cost for vibrating concrete is included in the costs for pouring concrete.

8.1.20 Table Saw

This item covers the rental charge for a table saw when required. The cost of the saw man is included in the cost of those activities requiring use of the table saw.

8.1.21 Finishing Machine

This item covers the rental charges for the rotary floating and troweling machine used on building slabs and the deck-finishing machine used on bridge decks. Deck-finishing machines are normally contractor owned. In our estimates we use a rental rate of $0.05 per SF of bridge deck to cover the cost of the deck-finishing machine, including fuel and maintenance. The cost of the operator is included in the labor cost to pour the bridge deck.

8.1.22 Develop Power

This item covers the cost to develop power for use on the concrete work. On building work, it normally includes the cost of a job service pole, the connect and disconnect charges by the power company, and the cost of additional electric cords if necessary. Portable generators are often used on engineering work since the power is required at a number of locations and often for only short periods of time. In such cases a monthly rental charge is made for the generators required, and this rental charge includes the cost of fuel and of additional electric cords when required.

8.1.23 Develop Water

This item covers the costs to develop the water required for the concrete item and includes the costs of the hoses and nozzles required to wet down the forms, cure the concrete, and water cut laitance from fresh construction joints when required.

8.1.24 Hoppers and Buggies

This item covers the rental charges for concrete hoppers, elephant trunks, and concrete buggies when required.

8.1.25 Neoprene-Bearing Pads and Expansion Joint Material

This item covers the cost, FOB job, to furnish neoprene-bearing pads used on certain bridges and expansion joint material.

8.1.26 Curing Paper

This item covers the cost, FOB job, for curing paper used on interior building slabs. On such work, paper is required to protect the slab even though curing compound is used. When taking off the paper required, an allowance must be made for overlaps at the sides. The labor cost to apply such paper on building slabs is included in the costs to pour and cure the slabs.

8.1.27 Curing Blankets

This item covers the cost, FOB job, to furnish the curing blankets, burlap, or polyethylene required to water cure structural concrete. Used rags are often used to water cure bridge decks and are sold by the ton. The labor cost to place and move this material is included in the cost to pour and cure the concrete.

8.1.28 Miscellaneous Iron

This item covers the cost, FOB job, of the miscellaneous iron required for the contractor's own use in connection with the formwork. Such material is not covered in the structural steel or miscellaneous iron items for the job. For example, the tie rods used to secure out-riggers to the exterior girders in connection with the formwork for deck overhangs is included here. The labor cost for installing the miscellaneous iron is included with the cost of the formwork involved.

8.1.29 Deck and Screed Hardware

This item covers the cost, FOB job, of screed pipes and supports on building work and of the deck form hangers used on precast concrete or structural steel girder bridges. These costs could be included with the miscellaneous iron but are listed separately to ensure that they are not overlooked.

8.1.30 Box Girder Dobies

This item covers the cost of the labor and material required to make the "dobies" used to support the interior girder and diaphragm forms on box girder bridges. These are concrete blocks $3\frac{1}{2}$ inches square with a height equal to the depth of the bottom slab of the box girder bridge. In our estimates we use a labor cost to make the dobies of $700 per MCY of concrete in the bottom slab and girder pour based on $K_{6+2} = 1000$. In addition, we allow 1.5 CY of concrete per MCY of concrete in the bottom slab and girder pour for the material required.

8.1.31 Clean Girders

When bridge decks are poured on steel girder bridges, the primed girder webs are inevitably streaked with water and cement. This streaking must be removed prior to applying the final coats of paint. In our estimates, we allow a labor cost of $120 per MSF of girder web to be cleaned based on $K_L = 13.48$.

8.1.32 Sweep Bridge Deck

Prior to obtaining final acceptance, the bridge decks must be swept clean. In our estimates we charge a labor cost of $13 per MSF of deck to be swept based on K_L = 13.48.

8.1.33 Clean Up and Out

This item covers the cost of moving all form materials off the job on completion. In our estimates we charge a labor cost of $43 per MFBM or MSF of form lumber or plywood ordered for the job based on K_L = 13.48. In addition, we charge a rental cost for trucks and for hoisting equipment if needed.

8.1.34 On-and-Off Charges for Steel Falsework

The costs of moving steel falsework on and off the job must be included when this material is rented from outside sources. In our estimates we assume that the steel falsework is contractor owned and that the cost of moving it onto the job is included in the rental charged. We charge the cost to move the steel falsework off the job on completion. Based on K_L = 13.48, we charge $34 per ton for labor and a rental cost for trucks and for hoisting equipment if needed.

8.1.35 Other

This item will cover the costs of any requirements relating to the concrete item which are not covered in the costs listed in the preceding. For example, the costs for mix designs or testing, if required, would be included here.

8.1.36 Retaining Walls

We now prepare a complete bid for the retaining walls shown in Figure 8.2. Reinforcing steel is a subcontract item, and no takeoff for it is necessary. For clarity, reinforcing steel is not shown on this, nor on succeeding plans.

We assume the job plans call for two retaining walls, one as shown in Figure 8.2 and one opposite hand. The two walls are to be constructed opposite each other 50 feet to the right and left of the centerline of a future roadway. The engineer's estimate for the retaining walls is as follows:

ENGINEER'S ESTIMATE

Item No.	Item	Unit of Measure	Estimated Quantity
1	Clearing and grubbing	LS	Lump sum
2	Structure excavation	CY	550
3	Structure concrete	CY	425
4	Reinforcing steel	LB	44000

Figure 8.2. Retaining walls.

The following job conditions are assumed:

1. Job is located 15 miles from the contractor's yard and from a rental crane source.
2. Clearing and grubbing consists of 1 tree 10 inches in diameter by 30 feet in height.
3. The earth is sandy loam.
4. There are no overhead obstructions.
5. A water supply is available at a nearby residence.
6. No traffic control is required.

Figure 8.3. Retaining walls, Truxtun Avenue Overpass, Contractor: Tumblin Company.

8.1.37 Structure Concrete

The structure concrete is the major job item; it is taken off and its cost is estimated first.

8.1.37.1 THE CONCRETE TAKE-OFF. The concrete takeoff is shown in Figures 8.4 and 8.5. The foundations and the walls are taken off separately since they represent two different operations. The primary activities required appear in abbreviated form as the column headings, and their quantities are taken off and entered in the appropriate columns.

The S_1 totals summarize the foundation work and show the CY of concrete, the contact area of foundation forms 1.5 feet in height, the SF to be fine graded, and the CY of hand excavation required for the slab key. The area to be fine graded is the contact area of all concrete with the subgrade. The miscellaneous column is used to collect additional items that may be unrelated to each other but which must be included in the estimate. In this instance, we note the gallons of curing compound required to cure the top face of the foundation slab.

Similarly, the S_2 totals summarize the wall work and show the CY of concrete and the contact areas of wall forms to be set and stripped. The wall areas are segregated by height and include the areas of the end bulkheads. The area to be sandblasted, the required amount of expansion joint material and chamfer strip, and the number of weep holes are noted in the miscellaneous column.

The total quantity of concrete in the retaining walls is found to be 423.0 CY, which is in reasonable agreement with the 425 CY shown on the engineer's estimate.

RETAINING WALLS	MISC.	H EX	FG	H15	I. CY
FOUNDATIONS					
$13.25 \times 30 \times 1.67/27$			398		24.6
74.5×1.67				125	
$2.00 \times 1 \times 30/27$		2.2			2.2
$12.00 \times 30 \times 1.50/27$			360	92	20.0
61×1.50					
$2.00 \times 1 \times 30/27$		2.2			2.2
$11.00 \times 30 \times 1.33/27$			330		16.3
61×1.33		2.2		82	2.2
$10.00 \times 30 \times 1.33/27$			300		14.8
70×1.33		2.2		93	2.3
1 ADD'L. FND.		8.8	1388	392	84.6
CURE COMP. $2776/150 = 19$ GAL.	1				
S_1	1	17.6	2776	784	169.2

WALLS	MISC.	H 17	H 19	H 21	H 23	CY
$A_1 = (1 + 24 \times 0.021) 24 = 36.1$					36	
$A_2 = (1 + 22 \times 0.021) 22 = 32.2$					33	
$68.3/2 \times 30/27$						37.9
$23 \times 2 \times 30$					1380	
$A_3 = (1 + 20 \times 0.021) 20 = 28.4$				29		
$60.6/2 \times 30/27$						33.7
$21 \times 2 \times 30$				1260		
$A_4 = (1 + 18 \times 0.021) 18 = 24.8$			25			
$53.2/2 \times 30/27$						29.6
$19 \times 2 \times 30$			1140			
$A_5 = (1 + 16 \times 0.021) 16 = 21.4$		22				
$46.2/2 \times 30/27$						25.7
$17 \times 2 \times 30$		1020				
4 WALLS ADD'L.		1042	1165	1289	1449	126.9
SAND BLAST $1.83 \times 240 = 440$ SF	1					
$\frac{1}{2}$" EJ MAT. $20 \times 2 \times 6 = 240$ SF	1					
CHAMFER $(20+1+20) 16$						
$\qquad +4 \times 120 = 1136$ LF	1					
WEEP HOLES 16 EA. 4"$\phi \times 2'$	1					
S_2	4	2084	2330	2578	2898	253.8
$\sum_1^2 S = 423.0$ CY						

Figure 8.4. Concrete takeoff for retaining walls.

The plans indicate a starter wall at the construction joint. It is not necessary to consider this in the estimate since the costs to set and strip starter walls and solid keys are included in the unit cost to set and strip foundations.

The takeoff for the general cost items is based on the quantities listed in sections S_1 and S_2. We decide to furnish foundation forms for one foundation complete and

RETAINING WALLS – CONT'D.	MISC.	Σ SF	SF MAKE	MSF PLY	MFBM 2×4	2. MFBM 2X HAVE
GENERAL COSTS						
S_1 FURNISH 400 SF STAKES & BRACES		784				0.8 0.3
S_2 MAKE 1165+1289+1449=3903 3.90 × 1.17 3.90 × 2		9890	3903	4.6	7.8	
GRIND 18 ×120×2 = 4320 SF	1					
HOLES 9600/4 = 2400 TOT. EXPOSED $\frac{2400}{2}$ - $\frac{240}{2}$ = 1080 }	1					
TAPER TIES $3903/8$ = 488 EA.	1					
FORM OIL $10674/150$ =71 GAL.	1					
S_0	4	10674	3903	4.6	7.8	1.1
				HAVE 2.3		

Figure 8.5. *Concrete takeoff for retaining walls (continued).*

to start setting wall forms the day after the first foundation pour. On this basis, wall forms must be furnished for the three larger wall pours, for one set of walls, to keep the crew from running out of work.

The foundation forms will be made of 2-inch material from the contractor's yard. The area from S_1 is converted to MFBM and is listed as shown together with an allowance for 2×4 stakes and braces. The unit costs for setting and stripping foundation forms include the cost of building the forms in place with 2-inch material; no additional allowance for making these forms is necessary.

The area of the wall form panels to be made is taken from section S_2 and is listed in the "SF Make" column. This area is converted to MFBM of 2×4 and MSF of plyform, as shown.

Since we plan to bid the additional finishing cost on a unit cost basis, the total contact area of forms is taken from S_1 and S_2 and is listed in the "Σ SF" column. The area to be ground is based on an assumed requirement of the specifications and includes all the exposed front wall areas above elevation 102.

The wall form ties are assumed to be set on 2-foot centers, and the number of form tie holes to be filled is determined by dividing the total area of formed wall faces from section S_2 by 4 SF per hole. This number is divided into exposed and unexposed holes since the costs differ. For this calculation it is assumed that on completion the bottom row of holes on the front wall faces will be covered.

The number of taper ties required is found by dividing the area of wall form panels to be made by 8 SF per tie. Taper ties are completely reusable. If snap ties or she-bolts were to be used, an allowance would be made for the cost of the embedded and lost parts, and this allowance would be based on the total area of all the walls.

A note is made at the bottom of the "MSF Ply" column that we have 2.3 MSF of used plyform in our yard.

8.1.37.2 THE STRUCTURE CONCRETE ESTIMATE.

The estimate sheet for the concrete item is shown in Figure 8.6. In all cases the applicable wage and rental rates are listed at the top of the sheet. Those items from the general cost checklist that are applicable are listed under the S_0 heading. The S_1 and S_2 items are listed directly from the concrete takeoff.

The unit costs relating to sections S_1 and S_2, together with the indexes on which they are based, are given in Table 8.3 and Figure 8.7. The unit costs used for the additional finishing items are shown in Table 8.2. The estimated cost of doing the work involved in sections S_1 and S_2 is determined, and an estimate of the time of completion for the item is made.

The estimated time to complete each section is determined in the days column of the estimate sheet. The total labor cost less the pour labor cost for section S_1 and for section S_2 is divided by the daily cost of the proposed crew to obtain the required number of working days to set and strip the formwork. A four-carpenter crew is assumed for the foundations, and a six-carpenter crew is assumed for the walls.

The cost of the pour labor is not included since only the first foundation pour and the last wall pour are on the critical path for the item. On larger jobs, the concrete work is normally completed well before the end of the job, and all pour times are neglected; those that are on the critical path for the item will be small compared with the times required for the total formwork and will be within the limits of accuracy for the time estimate.

In our case, the total time required is small, and the concrete work is essentially the entire job. Therefore, 1 working day is added to the foundation time to allow

```
ITEM 3.   425 CY STRUCTURE CONC.
          (423.0)
```

C 16.34 K6+2 1000			
L 13.48 Kcr 75	L	DAYS	M,R
F 16.10			

			L	DAYS	M,R
S0 GENERAL COSTS			{		{
CONC. 423.0 ×1.02 = 432 CY		30	}		12960
LUMBER BUY 7.8 MFBM		300	}		2340
✓ HAVE 1.1 ✓		100	}		110
PLY 2.3 MFS @ 500 + 2.3 MSF @ 170			}		1540
TIES 4.9 C 1 MO.		40c	}		195
MINOR EXP. 0.423 MCY		1950	}		825
MAKE FORMS 3900 SF		0.22	860		—
ADD'L. FINISH. 0.423 MCY		5980	2530		100
✓ CURE 5 WKS.		50	250		—
SAND BLAST 440 SF		0.30	130		20
COMPRESSOR 1 MO.		400	}		400
VIBRATORS (2) 1 MO.		200	}		200
GENERATOR 1 MO.		250	}		250
DEVELOP WATER 50+100			}		150
CJ MAT'L. 240 SF (½")	0.08	0.50	20		120
BURLAP 4 MSF			—		100
CLEAN UP 13.5 MFBM	43	12	580		160
S1 FOUNDATIONS			}	2130	}
18 CY H. EX.		14	250	1000×⅔	}
2780 SF FG		0.14	390	=4,+1	}
SS 784 SF H15		1.90	1490	5 D	}
1692 POUR 173 CY		6.92	1195		}
S2 WALLS			{		{
WEEP HOLES 16 EA @ 8 32 LF @ 2			130	12520	65
SS 2084 SF H17		1.20	2500	1000	
SS 2330 ✓ H19		1.23	2870	=13,+8	
SS 2578 ✓ H21		1.26	3250	21 D	
SS 2898 ✓ H23		1.30	3770		
2531 POUR 259 CY	6.10	5.90	1580		1530
423.0			21795		21065
			26045		

```
                    INCL. PAY ROLL EXP (1.195)
      Σ = $ 47,110    ($ 111/CY)                3+21
      ─────────────                             =24 D


      MINOR EXP.

            NAILS  10.5×0.423 = 4.4 CWT    40 |
            HAND TOOLS  0.423 MCY          500 |
            ELEC. TOOLS  1 MO.             70  |  $ 825
            FILING  5 WKS                 17  |
            CURE COMP.  19 GAL @ 1, +40        |
            FORM OIL    71 GAL @ 0.50, +40     |
            MILLWORK   1136 LF            0.13  |

      ADD'L FIN.

            TOUCH UP  10.67 MSF   30
            HOLES     1.08 M      600    $2530
                      1.32 M      300
            GRIND     4.32 MSF    270
```

Figure 8.6. Estimate for retaining wall concrete.

for the first foundation pour, and 8 working days are added to the wall time; 1 day is allowed for the last wall pour, 2 days are allowed for curing with forms in place, and 5 days are allowed for completing the cure and for the additional finishing. It should be remembered that the time required to strip forms is included in the time required for the formwork and that there are a minimum of 7 curing or calendar days for each 5 working days.

TABLE 8.3 Unit Costs for Retaining Walls

Activity	Labor Unit Cost ($)	Index	Rental Unit Cost ($)	Index	Unit of Measure
Foundations					
Fine grade	0.14	K_L = 13.48	—		SF
Hand excavation	14.00	K_L = 13.48	—		CY
SS H1–H3	1.90	K_{6+2} = 1000	—		SF
Pour foundations	3.72	K_L = 13.48	3.70	K_{Cr} = 75	CY
	6.92	K_L = 13.48	—		CY
Walls					
Make form panels	0.22	K_{6+2} = 1000	—		SF
Pour walls	6.10	K_L = 13.48	5.90	K_{Cr} = 75	CY

Note: When pours are large or average in size, the unit costs for the crane include 1 hour of travel time. On small pours, the travel time is additional, and travel time in excess of 1 hour should be added in all cases. Unit costs to set and strip retaining walls are given in Figure 8.7

Since half the foundation work will be overlapped with the wall work and all of the wall work is on the critical path, the total time to complete the concrete work will be 3 + 21 = 24 working days.

With the time for completion established, the items previously listed in the general costs section are priced. In practice, all the material and outside rental costs are based on quotations from the suppliers or on recent invoices. We have assumed that one of the laborers will monitor the weekend cure for $50 per weekend, that

Figure 8.7. *Unit labor costs to set and strip retaining wall forms when total area = 5000 SF or more. Increase unit costs 25 percent for areas of 1000 SF.*

Figure 8.8. *Retaining walls and steel beam bridge, Chester Avenue Underpass. Contractor: Tumblin Company.*

$50 will buy the water from the home owner, and that an additional $100 will cover the costs of hoses and nozzles. Although the cost to place incidental amounts of expansion joint material is included in the costs of the formwork, we have allowed an additional $0.08 per square foot to place this material because an unusual amount of layout is required.

The cost columns for labor and for equipment rental are totaled, and an allowance of 19.5 percent is added to the total labor cost for payroll expenses. This percentage is based on workmen's compensation insurance rates for concrete work NOC, and the method of determining the allowance for payroll expenses is discussed in Chapter 3.

The total cost for the item is entered on the estimate summary sheet illustrated later in Figure 8.10.

8.1.38 Structure Excavation

We assume that the specifications permit the structure excavation to be disposed of at a nearby location having an average haul distance of 250 feet one way. We also assume that the horizontal pay limits for structure excavation are specified to be vertical planes 1 foot outside the neat lines of the footings. The estimate sheet for the structure excavation item is shown in Figure 8.9.

The pay quantity of structure excavation shown on the engineer's estimate is checked and found to be 570 CY, including the 18 CY of hand excavation that was taken off and priced in section S_1 of the concrete estimate. The 570 CY figure is written in parentheses under the engineer's estimated 550 CY at the top of the sheet,

ITEM 2. 550 CY STR. EX. (510)			
5 CY RT LOADER $60/HR.	L	DAYS	M,R
LOADER ON-OFF 2 HRS 60 782 CY (TRUE) 0.50 Σ = $ 510	— —	$\frac{390}{60 \times 8}$ = 1 D	120 390 510
LOAD & DUMP 40 SEC 0.67 MIN. TRAVEL 500 LF/370 FPM 1.35 2.02 MIN. 55/2.02 × 5 × 0.9 = 122 CY/HR 782/122 = 6.4 SAY 6.5 HRS. 6.5 × 60/782 = $ 0.50/CY			

DIRECT JOB OVERHEAD
 	
 BONDS   70 M @ 7.50
 SUPT.   6 WKS @ 550
 RADIO   1.5 MO. @ 70
 TOILET   ✓   @ 60
 ICE & CUPS
 VEHICLES (2)   3 V. MOS. @ 280
 	
          INCL. PAYROLL EXP. (1.195)
 	
 Σ = $ 5555
 	
 TIME EST.
   .MOVE ON-OFF   2 D
     STR. EX   1
     CONCRETE   24
     REINF   3
          30 WORKING DAYS
	3300
	
	 3300
	 3945
			—
			 —
			 525
			 —
			 105
			 90
			 50
			 840
			 1610

Figure 8.9. *Estimates for structure excavation and direct job overhead.*

and we note that the item should overrun the engineer's estimate. The actual structure excavation based on 1 : 1 side slopes is found to be 782 CY. For this calculation the bottoms of the trapezoidal cross sections used are approximately the width of the concrete footing plus 1 foot on each side. The estimate for structure excavation is based on using a wheel loader with a 5-CY bucket. The calculation of the loader time required is shown on the estimate sheet. The loading and dumping time of 40 seconds takes into account the shallow cut and the need to make a rough grade 0.1 foot above finish grade. The average travel speed for the loader is taken at 370 feet per minute, and a 55-minute hour is used to compensate for minor delays. In addition, it is assumed that the average bank yards carried per bucket equal 0.9 of its rated capacity of 5 CY. A detailed discussion of excavation with wheel loaders is included in Chapter 4.

Since the structure excavation will be completed in 1 day at the start of the job, we assume that the required grade checking will be done by the superintendent, and no additional charge is made for grade-checking labor.

8.1.39 Clearing and Grubbing

Clearing and grubbing costs are discussed in Section 6.3. The item is included here to illustrate how the contractor's direct job overhead and his markup are distributed among the unit prices bid for the work.

For this job we assume that a local tree service will dispose of the tree for $80 after we have uprooted it with the loader used for the structure excavation, and we allow $\frac{1}{2}$ hour additional loader time for this. Since only two figures are involved, the cost information is entered on the summary sheet in the double space below the listed item.

8.1.40 Direct Job Overhead

The direct job overhead estimate is also shown in Figure 8.9. The items listed are taken from the direct job overhead checklist presented in Chapter 2. The bond cost is based on an approximate estimate of $70,000 for the bid price of the job, and the percentage used for payroll expenses applicable to the superintendent's salary is determined by the methods set forth in Chapter 2. The time estimate for the job is shown at the bottom of the direct job overhead sheet and is used to determine the cost of the superintendent.

8.1.41 The Estimate Summary Sheet

The estimate summary sheet is shown in Figure 8.10. The items from the engineer's estimate are listed, and the quantities are checked against those shown on the bid proposal. The costs for the contractor's own work including his direct job overhead are taken from the estimate sheets and listed in the cost column as shown. The unit price quoted by the low reinforcing steel subcontractor is extended and entered in the cost column. The column is totaled, 10 percent is added for general overhead and profit, and the total job cost is obtained. The cost previously used to calculate the bond cost on the direct overhead sheet is reviewed and found to be correct.

It is now necessary to assign unit prices to the various items in such a manner that all the direct job overhead and all the allowance for overhead and profit will be paid on completion and the total of the price column will be in reasonable agreement with the total of the cost column.

			COST		UNIT PRICE	TOTAL PRICE
1	CLEAR AND GRUB 80+30 = 110	LS	110		3000	3000
2	STRUCTURE EX.	550 CY	510		2.50	1375
3	STRUCTURE CONC.	425 CY	47110		130	55250
4	REINFORCING STEEL ACME 22.15/CWT DIRECT JOB OVERHEAD	44000 LB	9750 5560 63040		0.22	9680
		+ 10%	6300 69340			69305

Figure 8.10. Estimate summary sheet for retaining walls.

Reinforcing steel is a subcontract item, and we are not sure that the quantity shown on the engineer's estimate is correct. We therefore list a unit price for reinforcing steel equal to the low bid received rounded off to even cents. Thus, even though the item might underrun by a substantial amount, we will not lose any of our overhead and profit costs on the item.

We arbitrarily list $3000 as the lump sum price for clear and grub. This is less than 5 percent of the total job cost and is assumed to be in line with past bids for similar work in the area. The price of structure excavation is arbitrarily set at $3 per CY since we expect the item to overrun the engineer's estimate.

A trial determination of that unit price for concrete that will exactly balance the price bid with the total cost is made, and the balancing price is found to be $129.44 per CY. Since we prefer to bid our major item to an even dollar, we enter a unit price of $130 per CY for structure concrete and decrease the unit price bid for structure excavation to $2.50 per CY.

If the quantity shown on the engineer's estimate for structure concrete had been overstated by an appreciable amount, we would have been forced to bid the item at cost, which in this case would be $111 per CY, and to have balanced the cost bid prices by increasing the prices bid for clear and grub and for structure excavation.

It should be noted that regardless of the price we bid for subcontract items, we pay the subcontractors at the prices they quote to us in all cases.

8.1.42 Flat Slab Bridges

A cost estimate will now be prepared for the bridge shown in Figure 8.11. This bridge is assumed to be a portion of a larger project, and cost estimates will be illustrated for the concrete and the metal beam bridge rail items. The items for structure excavation, structure backfill, and reinforcing steel will not be considered. Excavation and backfill are discussed in Chapter 5, and reinforcing steel is a subcontract item in all cases.

It is assumed that the job site inspection reveals conditions similar to those noted in the previous estimates and that the water table is known to be below the proposed tip elevation for the piles.

8.1.43 Structure Concrete

It is assumed that the flat slab bridge (see Figure 8.12) constitutes the total concrete item for the job. This assumption is necessary because the items in the general costs section of the estimate for the concrete depend upon the total structure concrete involved in the project.

8.1.43.1 THE CONCRETE TAKEOFF. The concrete takeoff is shown in Figures 8.13 and 8.14 and includes a takeoff of the falsework lumber required. The falsework lumber is based on a preliminary design and is illustrated in Figure 8.15. In practice, the falsework lumber takeoff would be based on sketches made on the design sheets. Our takeoff assumes that the contractor has all of the falsework lumber except the 2 × 4's in his yard.

The form lumber is taken off in the general costs section S_0 and is based on areas from sections S_6 and S_7. The falsework lumber is taken from section S_8. It is decided to form and pour the abutments one at a time. Forms are made for one abutment plus one additional back face since the back face forms will be left in place until

Figure 8.11. Flat slab bridge.

196

FLAT SLAB BRIDGE			1.
	FG	HG	CY
ABUTM'TS & WINGWALLS			
2×4.75 x 36/27	72		12.7
34×6 + 36×4.75 + 2×2×4.75		394	
$\overline{4}$ ×14 x 1/27 × 2			4.2
($\overline{4}$ ×14×2 + 2×1 + 1.25×1) 2	56	231	
1 ADD'L ABUTMENT COMPL.	128	625	16.9
S_6	256	1250	33.8

	MISC.	CAP BMS.	DE	SOF	CY
DECK SLAB					
100×36 x 1.25/27					166.7
96×36				3456	
2×1.25×100 + 6×2.5×0.75			251		
3×2.5×0.75 × 36/27					7.5
6×0.75×36		162			
SET 42 SETS POST BOLTS	1				
S_7	1	162	251	3456	174.2

$\Sigma_6^7 S = 208.0$ CY

		CAP BMS. PADS	6×6 DE 4×4,6	SOF 6×12	CY 2X
FALSEWORK LUMBER $H_c 7$					
END BENTS 4×6 8/10		18	0.16		
TYP. BENT CAP 6×6 2/20			0.12		
POSTS 4×6 RO 9/6		9	0.11		
BRAC'G. 2×4 8/12					0.07
9 ADD'L BENTS		81	2.07		0.63
LONGIT. BRAC'G (5 LINES)					
2×4 30/8 70/12					0.72
STRINGERS					
6×12 72/12				5.2	
X BMS 4×4 102/5			0.68		
2×3 BLOCK 102/1					0.05
4×4 68/10 ×2 SPANS		=1.7	1.81		
4×4 72/10 × ✓		MFBM	1.92		
HAVE 13.8 MFBM S_8		108	6.9	5.2	1.5
BUY 1.5 ✓					

Figure 8.12. Flat Slab Bridge, Goose Lake Slough. Contractor: Tumblin Company.

the deck slab is poured. An allowance is made for reusing the abutment forms in the deck formwork.

8.1.43.2   THE STRUCTURE CONCRETE ESTIMATE. The estimate sheet for the structure concrete is shown in Figure 8.16, and the unit costs for doing this work are shown in Tables 8.11 (in Section 8.1.55.2 and 8.4 and in Figure 8.17.

FLAT SLAB BRIDGE (CONT'D)	MISC.	SF MAKE	MFBM HAVE	MSF PLY	2. MFBM 2 X
<u>GENERAL COSTS</u>					
S_6 MAKE $1250 \times \frac{1}{2} + 6 \times 34$		829			
$\quad$ 829 × 1.17, 1.75				1.0	1.4
S_7 413 × 1.17, 1.75		251		0.5	0.7
$\quad$ 3456 × 1.17				4.0	
RE-USE ABUTS					
$\quad$ (829 − 2×6×34) $\frac{2}{3}$ × 1.17, 1.75				(0.3)	(0.5)
S_8			13.8		1.5
FORM TIES $^{625}/_8$ = 78	1				
S_0	1	1080	13.8	5.2	3.1
				HAVE 4.0	

Figure 8.13. Concrete takeoff for flat slab bridge.

Figure 8.14. Concrete takeoff for flat slab bridge (continued).

The cost of minor expenses is determined by the procedure illustrated in Figure 8.6 since no recent cost data are available. The cost of 120 pairs of 2 × 4 wedges is included with the required chamfer strip under the millwork section. The allowance for the concrete buggy and collection hopper represents an expense that is actually related to the CIDH item see Chapter 7 but is included here because it appears on the checklist and is in this case a very minor item of expense.

The time estimate is made in the days column, and a two-carpenter crew is assumed for the abutment work. Six days are required to do the formwork, but only 3 days are on the critical path for the item. Three additional days are added to pour, cure, and backfill the front face of the last abutment, and the estimated critical path time to complete the abutments is 6 days. A six-carpenter crew is assumed for the falsework, and 5 days will be required. A four-carpenter crew is assumed for the deck slab work. Two days are required, and a total of 11 days are added, allowing 1 day to pour the slab, 8 working days to cure prior to stripping the falsework, and 2 days to complete the additional finishing. The total time required to complete the item is 24 working days.

8.1.44 Metal Beam Bridge Rail

The estimate for the bridge rail is also shown in Figure 8.18. The required anchor bolts, posts, rails, end sections, and rail bolts are normally furnished by a single-material supplier. The cost to set the anchor bolts for the posts was included in the cost to form the bridge deck slab.

The time required to set the rail is determined in the days column, and a two-carpenter crew will require 2 days to do this work.

Figure 8.15. Falsework for flat slab bridge.

ITEM 3. 208 CY STRUCTURE CONC.
(208.0)

C 16.34 K_{6+2} 1000
L 13.48
F 16.10 K_{cr} 75

		L	DAYS	M,R
S₀ GENERAL COSTS				—
CONC, 208.0 ×1.02 = 212 CY	30			6360
3.1 MFBM BUY	300			930
13.8 ″ HAVE	120			1655
1.2 MSF BUY	500			600
4.0 ″ HAVE	170			680
TIES 0.8C 0.5 MO.	40/c			15
MINOR EXP. 0.208 MCY	3000			625
MAKE FORMS 1080 SF	0.23	250		—
ADD'L FIN. 0.208 MCY	2750	570		20
ADD'L CURE 3 WKS.	50	150		—
SAND BLAST 0.208 MCY	390	80		15
COMPRESSOR 0.5 MO.	400			200
VIBRATORS 0.5 MO.	200			100
FIN. MACHINE 3600 SF	0.05			180
GENERATOR 0.5 MO	250			125
DEVELOP WATER 50+100				150
BUGGY + HOPPER				30
CURE RUGS				100
SWEEP 3.6 MSF	13	45		—
CLEAN. UP 16.9 MF.BM 43	12	725		200
S₆ ABUTM'TS			1990	
FG 256 SF	0.40	105	1000×⅔	
SS 1250 SF H6	1.51	1885	= 6	
POUR 35 CY (33.8)	4.35	150	3+3=6D	
S₈ FALSEWORK			4620	
MAKE PADS 1.7 MFBM	115	195	1000	
SS 108 PADS	7	755	= 5D	
SS 13.6 MFBM Hc7	270	3670		
S₇ DECK SLAB			1295	
SET 42 SETS AB	11	460	1000×⅔	
SS 162 SF BMS	0.87	140	= 2	
SS 251 SF DE	2.76	695	2+11=13D	
SS SOF. PLY INCL. W. FWK.		—		
POUR 178 CY 8.54 (174.2 / 208.0)	4.40	1520		785
		11395		12770
$\Sigma = \$27755$ ($133/cy) INCL. PR EXP. (1.315)		14985	Σ24D	

Figure 8.16. Estimate for flat slab bridge concrete.

TABLE 8.4 Unit Costs for Flat Slab Bridges

Activity	Labor Unit Cost ($)	Labor Index	Rental Unit Cost ($)	Rental Index	Unit of Measure
General costs					
Make form panels[a]	0.23	$K_{6+2} = 1000$	—		SF
Additional fishing	2750	$K_{L+F} = 29.58$	—		MCY
Sandblast[b]	390	$K_L = 13.48$	—		MCY
Abutments and wingwalls					
Fine grade	0.40	$K_L = 13.48$	—		SF
SS H5–6	1.51	$K_{6+2} = 1000$	—		SF
Pour	4.35	$K_L = 13.48$	—		CY
Deck slab					
SS deck edge	2.76	$K_{6+2} = 1000$	—		SF
SS pile cap side forms	0.87	$K_{6+2} = 1000$	—		SF
Pour[c]	8.54	$K_{L+F} = 29.58$	4.40	$K_{Cr} = 75$	CY
Set rail post anchor bolts	11.00	$K_{6+2} = 1000$	—		Set
Set posts and metal beam rail	2.38	$K_{6+2} = 1000$	—		LF
SS formed deck drains	8.15	$K_{6+2} = 1000$	—		Each

[a]Includes the costs to make abutment and deck edge forms.
[b]Includes the costs to sandblast the abutment construction joints and the tops of piles and pile extensions.
[c]If bridge deck finishing machine is not used, increase the unit cost for labor by $1.00 per CY.

Note: When pours are large or average in size, the unit costs fro the crane include 1 hour of travel time. On small pours, travel time will be additional.

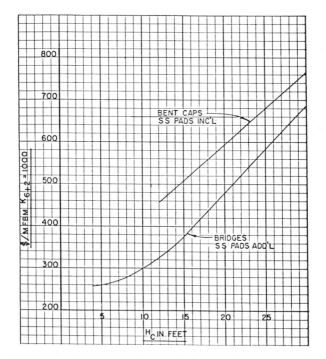

Figure 8.17. *Unit labor costs to set and strip wood falsework. Posts and caps 8 × 8 maximum, stringers 6 × 12 twelve-feet long maximum. The cost to set and strip plyform that is directly on the falsework is included.*

8.1.45 Total Time Required

The total time required to complete the flat slab bridge, including the CIDH piles and the bridge rail, will be 30 working days assuming the structure excavation and backfill are done by others and are not included. The drilling will be done concur-

Figure 8.18. *Estimate sheet for bridge rail.*

Figure 8.19. *T-Girder bridge with pile extensions, Kern River. Contractor: Tumblin Company.*

rently with the abutment work, the pile extensions will be done concurrently with the falsework, the railing will be set while the structure is curing, and the final cleanup can be done concurrently with the stripping of the falsework and the final additional finishing operation. Twenty-four working days are required to set, strip, and pour the bridge. One day is added to place reinforcing steel in one abutment, 3 working days are added to place the reinforcing steel in the deck slab, and 2 working days are allowed for moving on and off the job and to cover a possible delay resulting from the tight scheduling of the abutment and CIDH pile work.

8.1.46 Bridge Bents and Columns

Steel girder bridges (Figure 8.20) and prestressed concrete girder bridges are frequently supported on concrete bents such as the one shown in Figure 8.21.

8.1.47 The Concrete Takeoff

The concrete takeoff for the bent is shown in Figure 8.22. The general costs section of the takeoff is omitted since it depends on the total quantity of structure concrete on the job.

The required timber falsework is shown in Figure 8.23 and is based on a preliminary design. In practice, no formal falsework drawing would be required at the time the estimate was being prepared, and the falsework lumber takeoff would be based on sketches from the design sheets.

Steel shoring towers can be used to replace the posts, caps, and bracing shown in Figure 8.23 and, if used, will reduce the cost to set and strip the bents by a substantial amount. In this case, two three-frame, 1 B + E towers would be required (B + E means "base and extension").

Figure 8.20. *Bridge bent, plate girder bridge. Contractor: Tumblin Company.*

8.1.48 Unit Costs

The unit costs to do this work are shown in Figure 8.24 and Table 8.5. Unit costs shown in Figure 8.24 are applicable to the column forms and to the cap forms when the bent cap is separate from the girder system, as shown. On many T-girder and box girder bridges, the bent columns extend to the bottom of the girders, and the bent cap is replaced with a bent diaphragm which is formed and poured with the girders.

Table 8.4 gives unit costs for setting and stripping column footings poured in neat excavations as well as for setting and stripping footing forms when required.

Figure 8.21. *Concrete bent.*

CONCRETE BENT	MISC	FG	H3	CY
FOUNDATIONS				
$3(10 \times 10 \times \frac{3}{27})$		300	360	33.3
POUR NEAT $3(11 \times 11 \times \frac{3}{27}) - 33.3 = 7.0$ CY	1			
S_9	1	300	360	33.3
			H 14	CY
COLUMNS				
$3(4 \times 3 \times \frac{14}{27})$ S_{10}			588	18.7
		SOF	H5.5	CY
CAP				
$3 \times 5.5 \times 51.42/27$ S_{11}		119	599	31.4
$\Sigma_9^{11} S = 83.4$ CY				

	MISC.	4x12	4x6 6x6	2x
FALSEWORK				
1 BENT 6x6 $\frac{1}{5}$, 4x6 RO $\frac{2}{12}$			0.063	
2x4 $\frac{2}{5}$ $\frac{4}{8}$				0.028
5 BENTS ADD'L.			0.315	0.140
BRAC'G. BETW'N. BENTS				
2x4 $\frac{8}{12}$ $\frac{2}{8}$ $\frac{19}{16}$				0.245
STRINGERS 4x12 $\frac{4}{14}$ $\frac{8}{12}$		0.608		
X-B'MS 4x6 $\frac{36}{5}$ $\frac{8}{8}$			0.488	
12 PR. 2x4 WEDGES	1			
4 PADS 2' SQ = 0.1 MFBM				
TOTAL SS 1.9 MFBM H_c 14 S_{12}	1	0.6	0.9	0.4
ALL WOOD				
OR USE				
2 - ⋀⋀ 18 + E = 15.3 CWT				
PLUS 0.608 + 0.488 = 1.5 MFBM				

Figure 8.22. *Takeoff for concrete bent.*

Figure 8.23. Falsework for concrete bent.

SYM AB'T.
¢ BENT

5/8 PLY.

1-4×12 EA SIDE

4×6 12" C-C

2-4×12 EA SIDE

10'

10'

4'

6'

4×6 39/5 8/8

4×12
6×6 - 5'

4×6 R.O.

2-2×4 5'

8'

8'

4'

12'

2 @ 10'

8'

6'

8'

2 @ 10'

2×4 BRACING DIAGRAM

207

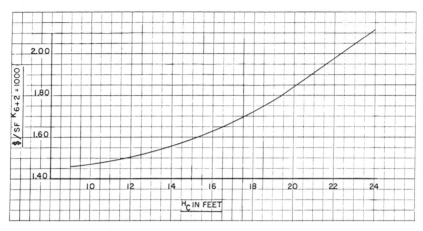

Figure 8.24. Unit labor costs to set and strip square or rectangular bent caps and columns.

When footings are poured against the earth, the costs given cover the setting and stripping of the column starter walls and the reinforcing steel templates. These costs are included in the costs of setting and stripping footing forms when side forms are used.

When foundation concrete is poured in neat excavations, considerable savings can be realized. The savings in formwork, excavation, and backfill costs will usually more than offset the cost of the additional concrete required.

In many sandy soils it is not practical to pour footings neat, and it is usually impractical when job conditions require the excavation to remain open for an extended period before pouring the concrete, or when the footings are small and relatively deep. Bridge column footings are suited to the neat pouring method, since they are regular in plan and can be poured quickly. In addition, their size is such that even if the banks slough in after the concrete has set up there will still be ample room to set and strip the columns.

Separate unit costs are given in Table 8.4 for pouring the columns and the bent caps. The columns must be poured before placing the reinforcing steel in the bent cap since the heavy cap reinforcing will prevent the use of elephant trunks during the column pour. Elephant trunks are normally specified to be used in all cases where the free fall exceeds 6 feet. Where the fall exceeds 6 feet and there is enough space to use elephant trunks, their use is mandatory to prevent segregation. Elephant trunks cannot be used when the clear distance between curtains of reinforcing steel is less than 11 or 12 inches. This situation occurs frequently on high wall pours. In such cases, segregation is prevented by the proper use of splash boards, which, in conjunction with the wall forms themselves, will prevent segregation provided the concrete is vibrated properly.

Wall pours as high as 28 feet are best poured by this method when the space available prevents the use of elephant trunks. Heavy-duty vibrators with long shafts

TABLE 8.5 Unit Costs for Concrete Bents and Columns

Activity	Labor Unit Cost ($)	Labor Index	Rental Unit Cost ($)	Rental Index	Unit of Measure
Foundations					
Fine grade	0.28	$K_L = 13.48$	—		SF
SS (neat excavation)	93	$K_{6+2} = 1000$	—		Each
SS side forms	1.80	$K_{6+2} = 1000$	—		SF
Pour foundations	4.45	$K_L = 13.48$	—		CY
Bents, wood forms[b]					
Make form panels	0.29	$K_{6+2} = 1000$	—		SF
Pour columns	8.30	$K_L = 13.48$	10.50	$K_{Cr} = 75$	CY
Pour caps	5.60	$K_L = 13.48$	7.80	$K_{Cr} = 75$	CY
Round columns, steel forms[b]					
SS	360	$K_{6+2} = 1000$	150	$K_{Cr} = 75$	Each
Pour	8.30	$K_L = 13.48$	10.50	$K_{Cr} = 75$	CY
Round columns, wood forms[c]					
Make forms	2.95	$K_{6+2} = 1000$	—		SF
SS	3.90	$K_{6+2} = 1000$	—		SF

[a]Columns and caps square or rectangular in cross section.
[b]Diameter 3–6 feet. Height 24–30 feet. On-and-off charges for crane are additional.
[c]Four feet in diameter by 12 feet in height.

Note: When pours are large or average in size, the unit costs for the crane include 1 hour of travel time. On small pours, travel time will be additional. Travel time for the crane used to set and strip steel column forms should be added in all cases. Unit costs to set and strip wood forms for rectangular or square bent columns and caps are shown in Figure 8.24

are required, as are experienced vibrator men. Attempts to pour such walls through pour windows in the wall forms will necessitate more lateral movement of the concrete with the vibrators, are costly, and will result in poor work in nearly every case. The splashboards normally used are made of plyform and are 22 inches wide by 8 feet long. They are suspended inside of the reinforcing steel curtains and are moved ahead as the pour progresses. Collection hoppers are used to direct the concrete between the boards.

Table 8.5 also gives unit costs for setting and stripping the round steel column forms frequently required in bridge construction. Although the unit costs for rental in connection with concrete work include a maximum of 1 hour on and off time for the crane, an exception is made in the case of setting and stripping steel column forms. For this work, the on-and-off charges for the crane must be added. This is necessary because only one or two sets of steel column forms may be available, and the number of units to which travel time would be distributed can easily vary by as much as 100 percent.

The costs shown for making, setting, and stripping wooden forms for round columns illustrate the advantages of using steel forms when large-diameter round columns are required.

8.1.49 Bridge Abutments and Wingwalls

Typical bridge abutments are illustrated in Figures 8.26 and 8.28. Bridge abutments are not designated by type numbers in practice, and the numbers shown are for this discussion only. Depending on the conditions, piles may be used with any of the abutments shown.

The type 1 abutment and wingwalls are similar to but deeper than the abutments used on the flat slab bridge previously discussed and are normally used on T-girder and box girder bridges. When such bridges are prestressed, a footing pad is used

Figure 8.25. *Bridge bents, steel beam bridges. Contractor: Tumblin Company.*

Figure 8.26. Bridge abutments.

211

Figure 8.27. Bridge abutments.

below the abutment wall shown. Expansion joint material is used at the construction joint between the pad and the wall to form a hinged joint, and the wingwalls are poured after the bridge has been post tensioned.

The type 2 and type 3 abutments are normally used on prestressed concrete girder and steel plate girder bridges. When type 2 abutments are used, the main foundation pad may return at the wingwalls for a short distance, and the small footing at the end of the wingwalls is eliminated. Type 4 wingwalls are usually used with type 3 abutments and are often built on stepped foundations.

Figure 8.28. Concrete barrier railing.

Unit costs for abutments are shown in Table 8.6. When stepped foundations are involved, the earth will slough in behind the steps, as was discussed in connection with the box culvert cutoff walls, and an allowance must be made for the estimated amounts of additional concrete and step forms required.

For estimating, the contact area of the step forms required is based on the height of the step at its front face and the distance between the foundation slabs at the sides. On type 2 abutments, the 8-inch walls at the sides are normally poured with the wingwalls and are considered as wingwalls when applying the unit costs shown.

In designing the forms for type 3 abutments, the vertical haunch load due to the weight of the concrete and the live load is distributed to the back wall form studs and to a beam and post arrangement at the rear edge of the haunch. The horizontal load from the fluid pressure of the concrete is held with form ties and wales that restrain the upper ends of the sloping haunch soffit joists. The unit costs shown for the setting and stripping type 3 abutments include the cost of setting and stripping the required beam and post system.

Small haunches involving increased wall thicknesses of 6 to 12 inches are sometimes required at the tops of walls. The forms for such haunches do not require the beam and post arrangement, and in such cases the unit price applicable to the wall without haunches is used to determine the setting and stripping cost, and the wall area used includes the actual contact area of the haunch form.

A comparison of the unit costs for retaining walls with those for type 4 abutment wingwalls shows that the abutment work costs more and the unit costs realized for the type 4 wingwalls do not vary appreciably with the height of the walls normally encountered. This clearly illustrates the need to recognize the subtle differences inherent in seemingly similar types of work, to develop separate unit costs where required, and to use sound judgment in applying unit costs to proposed work.

Retaining wall foundations tend to be more regular in plan than abutment work, they are simpler to lay out, and the steps, if any, are further apart. Larger areas of foundations and walls are normally involved in retaining wall work, and more efficient use of the crews is possible since a daily and repetitive sequence of operations can be established.

8.1.50 Barrier Railing

Typical bridge barrier railings are shown in Figure 8.28. Although barrier rails are designated by type numbers in practice, the designations vary, and the type numbers shown here are for this discussion only.

Usually 100 linear feet of concrete barrier are set and poured each day. The front barrier rail forms are stripped, and the exposed concrete is given a float finish on the same day the concrete is poured. Before acceptance, the back or outside face of the barrier and the concrete end posts, if any, are given a sacked finish.

Barrier railing is typically bid as a separate job item. In addition to the costs of setting, stripping, pouring, and finishing the concrete portion, the price bid includes the cost of the metal railing, which is furnished and set by a subcontractor. The cost of the reinforcing steel may be included in the price bid for the bridge-reinforc-

TABLE 8.6 Unit Costs for Abutments and Wingwalls

Activity	Labor Unit Cost ($)	Labor Index	Rental Unit Cost ($)	Rental Index	Unit of Measure
Foundation slabs					
Fine grade	0.28	$K_L = 13.48$	—		SF
Hand excavation	14.00	$K_L = 13.48$	—		CY
SS H1.5–H2.5 level	2.25	$K_{6+2} = 1000$	—		SF
SS H1.5–H2.5 stepped	2.60	$K_{6+2} = 1000$	—		SF
SS H5 steps	3.40	$K_{6+2} = 1000$	—		SF
Pour foundations	5.00	$K_L = 13.48$	—		CY
Abutments and wingwalls					
Fine grade type 1	0.40	$K_L = 13.48$	—		SF
Make form panels	0.24	$K_{6+2} = 1000$	—		SF
SS type 1, complete	1.95	$K_{6+2} = 1000$	—		SF
SS type 2 wingwalls	1.95	$K_{6+2} = 1000$	—		SF
Pour type 1, complete	5.25	$K_L = 13.48$	—		CY
Pour type 2 wingwalls	5.25	$K_L = 13.48$	—		CY
SS type 3 H12–H24	2.10	$K_{6+2} = 1000$	—		SF
SS type 4 H8–H18	1.55	$K_{6+2} = 1000$	—		SF
Pour types 3 or 4	5.85	$K_L = 13.48$	6.50	$K_{Cr} = 75$	CY
SS H4–H6 curtain walls	1.75	$K_{6+2} = 1000$	—		SF
Pour curtain walls	7.25	$K_L = 13.48$	—		CY
or	6.50	$K_L = 13.48$	11.00	$K_{Cr} = 75$	CY
SS bearing blocks	1.95	$K_{6+2} = 1000$	—		SF
SS curtain wall extensions	2.60	$K_{6+2} = 1000$	—		SF

Note: When pours are large or average in size, the unit costs for the crane include 1 hour of travel time. On small pours, travel time will be additional. Abutments and wingwalls, types 1–4, are shown in Figures 8.26 and 8.27.

ing steel or may be included in the price bid for the barrier railing, and the specifications must be checked regarding this in each instance. In all cases, the barrier-reinforcing steel is furnished and placed by the reinforcing steel subcontractor doing the bridge-reinforcing steel work.

Unit costs for barrier rails are shown in Table 8.7. The setting and stripping costs include the cost of setting the metal rail anchor bolts furnished by the railing subcontractor as well as of setting expansion joint material, as required. The material cost of the anchor bolts is always included in the price bid by the steel subcontractor who furnishes and erects the steel on them. The unit costs for pouring include the cost of the float finish on the front face.

The setting and stripping costs shown are applicable to steel or wood forms when based on cost per linear foot of barrier. When wooden forms are used with type 3 barrier rail, the walk portion is poured separately, whereas when steel forms are used, the walk is poured integrally with the wall.

The pay limit for barrier railing is the measured length of the concrete barrier and includes the length occupied by the end posts when concrete end posts are used. The additional formwork required for end post details is priced at the unit cost per square foot of contact area used for the barrier wall.

In practice, the barrier rail item is priced after the cost estimate for the bridges has been completed since some of the general cost items for the barrier rail are related to those used for the bridge work. We now present a complete cost estimate for a T-girder bridge with type 3 barrier rails.

8.1.51 T-Girder Bridges

The T-girder bridge is shown in Figure 8.29. Type 1 abutments with the dimensions shown in Figure 8.26 are used. It is assumed that the job conditions and location are the same as those assumed in the previous estimates.

8.1.52 Structure Concrete

It is assumed that the T-girder bridge constitutes the total structure concrete item for the project. We plan to pour the deck with the girders since considerable savings are realized by this method. The sandblasting otherwise required at the construction joint between the girders and the slab is eliminated, as is the cost of the after-pour cleanup and the additional curing expense required when the girders are poured separately.

8.1.52.1 THE CONCRETE TAKE-OFF. The proposed falsework and slab soffit forms are shown in Figures 8.30 and 8.31 and are based on a preliminary design. The concrete takeoff is shown in Figures 8.32 through 8.35 and includes the takeoffs for the falsework and soffit forms.

As previously mentioned, the falsework takeoff would normally be based on sketches made on the design sheets. It is of utmost importance, however, that the crews be furnished with good working drawings of the falsework before its erection. Good falsework drawings result in well-built falsework, and vice versa.

TABLE 8.7 Unit Costs for Concrete Barrier Railing

Activity	Labor Unit Cost ($)	Labor Index	Rental Unit Cost	Rental Index	Unit of Measure
Make wall forms	1.17	$K_{6+2} = 1000$	—		SF
Make wall face	0.25	$K_{6+2} = 1000$	—		SF
SS type 1	1.60	$K_{6+2} = 1000$	—		SF
	6.13	$K_{6+2} = 1000$	—		LF
SS type 2	6.13	$K_{6+2} = 1000$	—		LF
SS type 3 wall	1.60	$K_{6+2} = 1000$	—		SF
SS type 3 walk face	2.50	$K_{6+2} = 1000$	—		SF
SS type 3 wall plus walk, complete	9.27	$K_{6+2} = 1000$	—		LF
Pour type 1, 2, or 3 walls[a]	43.50	$K_{L+F} = 29.58$	—		CY
Pour type 3 walk	26	$K_{L+F} = 29.58$	—		CY

[a]Includes the cost to float finish the front face of the wall. The cost to sack the back face and the cost to sandblast the construction joint at the deck are additional.

Note: Barrier railings types 1, 2, and 3, are shown in Figure 8.28.

Figure 8.29. T-Girder bridge.

Figure 8.30. Falsework for T-girder bridge.

Figure 8.31. Falsework for T-girder bridge.

The limits for falsework and soffit formwork are shown in Figure 8.31. The girder form panels are shown as well since the soffit form ledgers are nailed to the girder form studs. The labor cost to set and strip plywood that is directly on the falsework is included in the unit costs to set and strip the falsework.

The relationship between the three types of formwork shown in Figure 8.31 must be thoroughly understood. The girder form panels are made, set, and stripped at the unit costs for this work on a square foot basis; the slab soffit forms are set and stripped at their unit costs on a square foot basis, and the falsework is set and stripped at unit costs based on the MFBM of lumber and the number of beams or the CWT of steel shoring required. In all cases, the cost of the materials is based on the MFBM of lumber and the CWT of steel required.

The concrete takeoff shown provides an orderly and systematic method for taking off and collecting the quantities required and will now be discussed in detail.

The type 1 abutments and wingwalls are taken off first. The S_{13} total indicates that 216 SF must be fine graded, 1436 SF of forms 5 to 10 feet in height must be set and stripped, and 42.8 CY of concrete are required.

The S_{14} total indicates that 200 SF must be fine graded, 180 SF of foundation forms 2.2 feet in height must be set and stripped, and 16.7 CY of concrete are required. The miscellaneous item notes the additional concrete required if the footings are poured in neat excavations, in which case the 180 SF of forms will not be needed.

The S_{15} total indicates that 560 SF of column forms must be set and stripped and 17.8 CY of concrete are required.

The S_{16} total summarizes the deck and girder work and shows the contact areas of the deck edge forms, girder soffits plus slab soffit, girder sides, and the amount of concrete involved. The approximate height of the deck edge forms and the girder

T GIRDER BRIDGE			FG	H 5/10	CY
ABUTMENTS & WINGWALLS					
$2.5 \times 5 \times 32/27$			80		14.8
$34.5 \times 2 \times 5$				.345	
$\overline{6.38} \times 14 \times 1/27 \times 2$			28		6.6
$(\overline{6.38} \times 14 \times 2 + 3 + 4.75) 2$				373	
ONE ADD'L. ABUTM'T.			108	718	21.4
		S_{13}	216	1436	42.8
	MISC.	FG	H $2\frac{1}{2}$	CY	
BENT F'ND'S.					
$2(10 \times 10 \times 2.25/27)$			200	180	16.7
IF POUR NEAT ADD'L CONC.= 3.5 CY	I				
$2(11 \times 11 \times 2.25/27) - 16.7$					
		S_{14}	I	200	180	16.7
				H20	CY
BENT COL'S. $2(4 \times 3 \times 20/27)$		S_{15}		560	17.8
	7"DE	SOF	H $4\frac{1}{2}$	CY	
GIRDERS + SLAB					
ABUT. DIAPHRAGM					
$2.5 \times 4.21 \times 32/27$					12.5
$(4.75 + 4.21) 32 + 2.5 \times 4.75 \times 2 - 4.21 \times 1.25 \times 5$				284	
I ADD'L DIA.				284	12.5
BENT DIAPHRAGM					
$4.46 \times 27.92 - 1.25 \times 0.25 \times 2 = 123.9$ SF					
$123.9 \times 3/27$					13.8
$123.9 \times 2 + 4.08 \times 3 \times 2 - 1.25 \times 4.21 \times 10$				220	
$(27.92 - 8) 3$			60^G		
GIRDERS L = 132 LF					
$132 \times 1.25 \times 4.21/27 \times 5$			825^G		128.6
$(3.83 + 4.21) \times 132 \times 2 + 4.21 \times 2 \times 132 \times 3$				5457	
INTERMED. DIA. $5.42 \times 3.71 \times 0.6/27 \times 8$				322	4.0
SLAB					
$32 \times 140 \times 0.542/27$					89.9
$(32 - 5 \times 1.25) 132 + 2.04 \times 3 \times 2$			3411^S		
ADD'L. @ OH					
$(\overline{0.75} - 0.54) 2.04 \times 135/27 \times 2$					4.3

Figure 8.32. Concrete takeoff for T-girder bridge.

side forms is indicated in the headings of these columns. The miscellaneous column notes the contact areas of deck and girder pour bulkheads since we plan to pour the combined deck and girder system in two pours. The total areas of girder and slab soffits are also noted in the miscellaneous column.

The falsework takeoff is summarized in the S_{17} totals, which indicate the CWT

T GIRDER BRIDGE (CONT'D)	MISC.	7"DE	SOF	H4½	2. CY
GIR'S. + SLAB (CONT'D)					
DECK EDGE 135 × 0.583 × 2		158			
POUR BULKHEADS					
DECK 0.54 × 32 + 2.04 × 0.21 × 2 = 18 SF	1				
GIR'S. 5 × 1.25 × 4.21 = 26.5 SF					
GIR. SOF 885 SF ⎫ 4296 SF	1				
SLAB ⌄ 3411 SF ⎭					
$\Sigma_{13}^{16} S = 342.9$ CY S₁₆	2	158	4296	6567	265.6

	MISC.	CWT 2B+E	MFBM SS	4×6	6×12
FALSEWORK					
BENT N1 2B+E 2 @ 9.45 CWT		18.9			
6×12 6/12 , 4×6 23/6			0.71	0.28	0.43
2×4 FILL 4/32			0.09	0.09	
GIR'S.					
NN 2B+E 10 @ 10.41 CWT		104.1			
N1 2B+E 10 @ 9.45 CWT		94.5			
6×12 80/12 (S1E)			5.76*		5.76*
16 WF 36 20/24 = 173 CWT	1				
STRUTS 4×6 60 LF			—	0.12	
RODS ON WORK SHEET	1				
END BENTS 4×6 68 LF			0.14	0.14	
FLAT PADS (4SF) 20 EA.	1		—	0.5	
4×6 X BEAMS					
14'⟍ 12' 14'					
16'1 16'1 16'1 16'1 16'1					
STR. SPA. 4' 5' 4' 2.7' 4' 2.7' 4' 5' 4					
L = 70 - 2.5 - 1.5 = 66' EA. SPAN					
67 + 1 @ DIA. = 68, × 2 = 136 LINES					
EVERY 5ᵀᴴ LINE CONT.					
136 × 1/5 = 28, +2+2 = 32 CONT.					
4×6 64/14 32/12 520/6			8.80	8.80	
* S1E S₁₇	3	218	15.5	9.7	6.2

Figure 8.33. *Concrete takeoff for T-girder bridge (continued).*

of 2 B + E steel shoring and the MFBM of lumber to be set and stripped as well as the MFBM of 4 × 6 and 6 × 12 to be furnished. The miscellaneous column indicates the required number of flat pads for use under the wide flange (WF) beams and the number of WF beams required and notes that the rod bracing for the top flanges of the beams will be taken off and priced on a separate work sheet.

T GIRDER BRIDGE (CONT'D)	3. 4×4 4×6	2×
SOFFIT FORM LUMBER		
DECK OVERHANGS L=135' = 27 @ 5' EA. SIDE		
SILL 4×6 $^{18}/_{16}$	0.58	
POSTS 4×4 $^{56}/_{4}$	0.30	
CAPS 4×4 $^{56}/_{4}$	0.30	
LEDGER + BRIDGING 2×4 $^{18}/_{16}$ $^{18}/_{16}$		0.38
BEAMS 4×6 $^{18}/_{16}$	0.58	
2×6 ✓		0.29
2×4 ✓		0.19
INTERIOR SOF. L=4 @ 32.67'		
1 BAY 2×4 $^{136}/_{6}$		0.55
2×6 $^{24}/_{12}$		0.29
3 ADD'L. BAYS		2.52
S$_{18}$	1.76	4.22

TOTAL SOF. $\dfrac{5.98}{3.41} = 1.75 \dfrac{MFBM}{MSF}$

OH SOF. $\dfrac{2.62}{0.551} = 4.76$ ✓

INT. SOF. $\dfrac{3.36}{2.86} = 1.17$ ✓

Figure 8.34. Concrete takeoff for T-girder bridge (continued).

The falsework takeoff assumes the column footing excavations will be backfilled before erecting the falsework. This is good practice because important savings are realized when bridge falsework is erected on clean and approximately level surfaces. Excavation and backfill are separate job items and are discussed in Chapter 5. With proper scheduling, the backfilling can be done outside the critical path for the project.

T GIRDER BRIDGE	MISC	MAKE	16WF36	2B+E	HAVE	PLY	4. 2X
GENERAL COSTS							
S₁₃ ABUT'S. MAKE 1 718 × 1.17, 1.75		718				0.84	1.26
S₁₄ BENT FND'S. POUR NEAT 3.5 CY	1						
S₁₅ COL'S. MAKE 2 560 × 1.17, 2.00		560				0.66	1.12
S₁₆ GIR'S + SLAB 6567 × 1.17, 1.75		6567				7.68	11.5
4296 × 1.17						5.03	
158 × 1.17, 1.75		158				0.18	0.28
RE-USE S₁₃, S₁₅ PLY 0.84×2/3 } 0.66 × 1 }						(1.22)	
2× (1.26+1.12) 2/3							(1.59)
S₁₇ FALSEWORK WF BRACING	1		20 EA =17.3 CWT	218	15.9		
S₁₈ SOF. FORMS					1.8		4.2
TAPER TIES 6559/8 =820	1						
SHE-BOLTS (20-2)/2 = 9 SPA 2 (10 ×5) = 100 BOLTS} 40 ½"φ × 46" } 60 ½"φ × 30 }	1						
So	4	8003	↑	218	17.7	(13.2	16.8
						HAVE 8.0	

Figure 8.35. *Concrete takeoff for T-girder bridge (continued).*

In contrast, the falsework for the concrete bents discussed previously was erected before backfilling the column footing excavations. In that case, savings could be realized by using the cap falsework for staging during the column pours, and the construction sequence would be interrupted if the the backfilling were done between the column and the bent cap operations.

The cost to set and strip the flat pads required for the steel shoring is included in the unit cost to set and strip the shoring, and we assume we have these pads in our yard with the shoring. The cost to set and strip the top flange bracing and the bridging at the ends of the beams is included in the unit costs to set and strip these beams. No allowance is made for buying the bridging material since we plan to use scrap material for this.

A precise determination of the top flange bracing required to reduce the *ld/bt* ratio is not possible. When T-girder bridge decks and girders are poured together, the total dead load of the concrete must be taken by the falsework at one time, and in view of the grave consequences of a falsework failure, we use the rod and strut system indicated for estimation. Before erecting the falsework, we investigate the bracing and beam system with respect to the horizontal loading involved.

S1E 6 × 12's are required for some of the falsework, but this poses no problem since wooden falsework beams are normally bought S1E. (surfaced one edge) In this case the S1E beam is 9 percent stronger than an S4S (surfaced four sides) beam, while the MFBM and the cost are the same for both.

The S_{18} total summarizes the soffit form lumber required and is broken down into 4 × 4 and 4 × 6 lumber and 2-inch-thick material.

The conversion factors in MFBM per MSF for the soffit formwork are noted at the bottom of the sheet for rough checking of future takeoffs. When slab soffit formwork constitutes only a small portion of the contractor's work, the material cost is often based on such factors since the labor cost is based on the square feet involved and is not affected by inaccuracies in the conversion factors used.

No allowance is made on the takeoff for the blocking or the short 2 × 4 post required at the deck overhangs. This material will be made from scrap lumber from the other formwork.

The general costs section is primarily a material summary and is based on the S_{13} through S_{18} totals. The contact areas of the required forms are converted to MSF of plyform and MFBM of 2-inch-thick (2X) lumber using the factors discussed in the general costs section of this chapter. In this case, we plan to make one abutment form and will reuse it on the second abutment. Two column forms will be made since the pours are small, only a small quantity of forms are involved, and the material will be reused in the soffit forms.

We assume we have used 4 × 4, 4 × 6, and 6 × 12 material in our yard, and the required MFBM of this material is summarized in the "have" column of section S_0. We must use 13.2 MSF of plyform, and we note that we have in our yard 8.0 MSF of used plyform that will be used for interior girder and soffit formwork.

The miscellaneous column notes that we will pour the column footings neat and that 3.5 additional CY of concrete are required. The note to take off the wide-flange bracing is made to ensure that this is not overlooked. The form ties needed are also taken off and noted in this column. Eight hundred twenty taper ties will be required for the girder form panels, and these same ties are ample for the abutment work.

Because of the lengths involved, she-bolts with $\frac{1}{2}$-inch-diameter inner tie rods will be used to tie the column forms. We will use three 36-inch and four 48-inch she-bolts for each 2 feet of height, and the bottom bolts will be located 1 foot above

Taper ties

Waler rod or she—bolt

Snap tie

Figure 8.36. Form ties (The Burke Company).

the bottom of the column form. Fifty she-bolts are required per column together with 20 tie rods 46 inches long and 30 tie rods 30 inches long.

The tie rod lengths are based on a rule of thumb that a 12-inch wall formed with 2 × 4 studs and double 2 × 4 wales will require a 2-inch tie rod. In this case, we estimate that 2 × 8 wales will be used with the 48-inch she-bolts and 2 × 6 wales will be used with the 36-inch she-bolts. When she-bolts are used, nearly all of the inner tie rods remain embedded in the concrete, and tie rods must be furnished for each use under normal circumstances. Although in our case no reuse is possible, the columns will be tied with two of the 36-inch bolts and both of the 48-inch bolts located just outside of the column form studs, and only one 30-inch tie rod would be embedded and lost for each set of five she-bolts used.

Typical she-bolts, taper ties, and snap ties are illustrated in Figure 8.36.

8.1.52.2 THE STRUCTURE CONCRETE ESTIMATE. The estimate sheets for the structure concrete are shown in Figure 8.37 and 8.38. Work sheets relating to some of the general costs items are shown in Figures 8.39 and 8.40. Unit costs for the girder and deck work are shown in Table 8.8, and the unit costs for the abutments, columns, and falsework have been discussed previously.

The miscellaneous iron item includes the costs for the WF beam bracing and column tie rods. The cost of a portable welding machine and a welder are included since the bracing for the beams will be fabricated on the job.

The minor expenses item is set up and estimated on the work sheet since this item consists primarily of material costs which cannot be indexed. Where recent cost histories for similar work are available, this item would be estimated as cost per MCY of concrete involved.

The number of working days required to complete the work summarized in each section is based on using a four-carpenter crew on the abutments and columns, a two-carpenter crew on the column foundations, and six-carpenter crews on the girders plus slab and on the falsework. Two working days are added to the girder plus slab work to cover the 2 pours required, and 8 working days are added to comply with an assumed requirement of the specifications for a 10-calendar-day curing period prior to commencing the stripping operations.

The number of working days required to complete the item is determined at the bottom of the estimate sheet and amounts to 45 working days. Since 8 of these days represent a curing period, 2 months are allowed for rental periods in the general costs section.

8.1.53 Barrier Railing

We assume the bridge has type 3-3 barrier railing on one side and type 2 barrier railing on the other. The 3-3 designation indicates the concrete walkway is 3 feet wide. The pay length for each railing will be 168 LF out-to-out of the concrete end posts. Details for concrete barrier rails were shown in Figure 8.28.

ITEM 46 340 CY STRUCTURE CONC.				
C 16.34 K$_{6+2}$ 1000 L 13.48 F 16.10 K$_{cr}$ 75		L	DAYS	M, R
S$_0$ GENERAL COSTS				
CONC. 342.9 × 1.02 = 350 CY	30			10500
16.8 MFBM 2 X	300			5040
17.7 ✓ HAVE	120			2120
5.2 MSF PLY	500			2600
8.0 ✓ HAVE	170			1360
218 CWT SHOR'G. 2 MOS.	436			870
173 ✓ WF 1 USE	1.50			260
8.2 C TIES 2 MOS.	40/c			655
1.0 C SHE-BOLTS 0.5 MO	✓			20
MINOR EXP. 0.343 MCY	3560			1220
MAKE 8003 SF	0.26	2080		—
ADD'L. FIN. 0.343 MCY	5180	1775		70
ADD'L. CURE 6 WKS.	90	540		—
SAND BLAST 271 SF	0.30	80		15
COMPRESSOR 2 MOS.	400			800
VIBRATORS ✓	200			400
DECK FIN. MACH. 4480 SF	0.05			225
GENERATOR 2 MOS.	250			500
DEVELOP WATER 100+150				250
HOPPERS				20
CURE RUGS				130
MISC. IRON				660
SWEEP 4.48 MSF	13	60		—
CLEAN UP 47.7 MFBM 43	12	2050		570
20 T 34	7	680		140
S$_{13}$ ABUT. + WW		—		
FG 216 SF	0.40	85	2885	
SS 1436 SF H $\frac{5}{10}$	1.95	2800	1000×2/₃	
POUR 44	5.25	230	=5 D	
S$_{14}$ BENT F'ND.		—		
FG 200	0.28	55	245	
3.5 CY ADD'L. CONC. 1	30	5	1000×1/₃	105
SS 2 F'ND'S.	93	185	=1 D	
POUR 17	4.45	75		

Figure 8.37. Estimate for T-girder bridge concrete.

ITEM 46 340 CY STR. CONC. (CONT'D.)

				L	DAYS	M,R
	S_{15} BENT COL'S.			}	1030.	}
	SS 560 H20		1.84	1030	1000 × $\frac{2}{3}$	
17.8	POUR 18 CY	8.30	10.50	150	= 2 D	190
	S_{16} GIR'S. + SLAB			}		
	SS DECK B'LKH'D. 18 SF		5.30	95	17315	
	SS GIR ✓ 26 SF		1.89	50	1000	
	SS 6567 H4$\frac{1}{2}$		1.89	12410	=18	
	SS 3411 SF SOF.		1.20	4090	+2+8	
	SS 158 SF DE H7"		4.25	670	=28D	
2656	POUR 271 CY	7.00	5.00	1900		1350
3429	S_{17} FALSEWORK			}		}
	SS 218 CWT 2B+E	8.90	0.60	1940	6150	130
	SS 20 FP		7	140	1000	—
	SS 20 16WF36 24'	40	29	800	=7 D	580
	SS 15.5 M.FBM	211	54	3270		835
				37245		31615
	INCL. PAYROLL EXP. (1.315)			48980		

$\Sigma = \$80,595$ ($\$235$/CY)

TIME EST.	CONSEC.	CONC'T.
ON - OFF	2	2
ABUT'S.	5	3
BENT FND'S.	1	1
COL'S.	2	2
GIR. + SLAB	28	28
F'W'K.	7	5
REINF.	—	}
ABUT	1 }	
COLS	1 }	4
GIR + SLAB	6 }	—
	53 D	45 D

Figure 8.38. Estimate for T-girder bridge concrete (continued).

ITEM 46   WORK SHEET

			M,R
MISC. IRON			
44 ℞ 4×⅜×4 1 HOLE EA.	3		132
16 ℞ 2½×⅜×12 ✓	4		64
80 LF ¾" φ ROD	0.40		32
8 ¾" TURNBUCKLES	7		56
36 ¾" φ × 9" MB	1		36
8 ¾" φ × 2" ✓	0.50		4
40 ½" φ × 46" TIE RODS	1.20		48
60 ½" φ × 30" ✓	0.80		48
PORTABLE WELDER 8 HRS.	30		240
			660
MINOR EXPENSES			
NAILS 0.343 MCY ⓐ 10.5=4 CWT	40		160
HAND TOOLS 0.343 MCY	500		170
ELECTRIC TOOLS 2 MOS.	70		140
FILING ✓	70		140
CURE. COMP (DECK) 30 GAL	1+40		70
FORM OIL 90 GAL	0.50+40		85
WEDGES 20 PR	0.40		8
CHAMFER 3440 LF	0.13		447
(0.343 MCY ⓐ 3557)			1220
SAND BLAST			
ABUT'S. (2.5×32+4.75×2) 2 ⎫			
COL'S. 3×4×4 ⎬ 271 SF			
POUR J'TS. 18+26 ⎭			

Figure 8.39. Work sheet for T-girder bridge estimate.

ITEM 46. WORK SHEET

20T FLAT BED TRUCK 30 K$_{cr}$ 75 FORK LIFT 30	L		M,R

ADD'L. FINISHING

GRIND COL'S. + WW 0.784MSF	270	212	
14X18X2+ 5̄ X14X4			
SACK OS GIR + DE 1.23 MSF	500	615	
140X3.83X2 + 158			
FILL TIE HOLES			
7995/4 + 40 (COL.) = 2039 TOT.			
OS GIR 3.83X132 X²/4=0.253 M	600	152	
OTHER 2039 - 253 = 1.786 M	300	536	
TOUCH UP 8.713 MSF	30	261	
(0.343 MCY @ 5178)		1776	

CLEAN UP RENTAL

TRUCKS 12 MFBM OR 20T STEEL PER LD.
  5 LDS. REQ'D.   USE 2 TRUCKS
  30M/40MPH = 0.75HR TRAVEL
               + 0.67 ✓ LOAD-UNLOAD
                 1.42 ✓ CYCLE SAY 1½HR
HAUL COST PER LD. = 1.5X30   **45**
CRANE   LOAD TIME 20MIN
  30M/30MPH =1.00 HR TRAVEL
  LOAD TIME 4 @ 0.75 HR + 0.33HR = 3.33HRS
  3.33 + 1.00 = 4.33 HRS   SAY 4.5 HR
  LOAD COST PER LOAD 4.5X75/5   **68**
FORK LIFT   UNLOAD IN 20MIN
  TRAVEL TIME ON-OFF ASSUME 0.67HR
  UNLOAD TIME = LOAD TIME   3.33HR
  UNLOAD COST PER LOAD 4X30/5   **24**
                                   137

    137/12 = $11.40 PER MFBM
    137/20 = $6.85 PER TON

Figure 8.40. Work sheet for T-girder bridge estimate (continued).

8.1.53.1 THE CONCRETE TAKEOFF. The takeoff for the type 3 railing is shown in Figure 8.41. The S_{19} totals show the cubic yards of concrete in the walkway and the wall section separately since two distinct operations requiring different unit costs are involved. In addition, the contact areas of the wall, walkway face, and end posts are listed.

TABLE 8.8   Unit Costs for T-Girder Bridges

	Labor		Rental		
Activity	Unit Cost ($)	Index	Unit Cost ($)	Index	Unit of Measure
Make girder form panels	0.26	$K_{6+2} = 1000$	—		SF
SS girder forms and pour bulkheads	1.89	$K_{6+2} = 1000$	—		SF
SS deck pour bulkheads[a]	5.30	$K_{6+2} = 1000$	—		SF
SS net slab soffit	1.20	$K_{6+2} = 1000$	—		SF
SS deck edge H6"–H7"	4.25	$K_{6+2} = 1000$	—		SF
Pour girders and slab	7.00	$K_{L+F} = 29.58$	5.00	$K_{Cr} = 75$	CY
Pour girders only	5.00	$K_{L} = 13.48$	4.80	$K_{Cr} = 75$	CY
Pour deck slab only	11.20	$K_{L+F} = 29.58$	7.05	$K_{Cr} = 75$	CY

[a]Increase unit cost 30 percent when bulkhead is skewed with respect to the transverse reinforcing steel.

Note: The equipment costs for pouring assume normal access for the crane, and the concrete can be deposited directly into the forms. When pours are large or average in size, the unit costs for the crane include 1 hour of travel time. On small pours, travel time will be additional.

As previously noted, when considerable amounts of barrier railing are required, 100 LF of railing wall is normally completed per day. In this case, we plan to make 82 LF of railing forms plus the forms for one end post and to cut off the wall forms for reuse on the type 2 railing.

For this illustration, the areas of the required forms are converted in the general costs section to MSF of plyform and MFBM of 2X material. We note that we have this material and will reuse material previously furnished for the bridge formwork.

The form tie holes to be filled consist of the holes in the end posts plus those on the back face of the wall. The holes on the front face will be closed at no additional cost during the float-finish operation.

An allowance is made in the minor expenses section for the use of driven studs at 4-foot centers to hold the forms to line.

The takeoff for the type 2 barrier railing is shown in Figure 8.42 and the area of the forms to be modified is noted in the general costs section. No form ties pass through the type 2 wall, and only those holes in the end posts require patching.

8.1.53.2   THE BARRIER RAILING ESTIMATES. The estimate sheets for the barrier railings are shown in Figures 8.43 and 8.44, and the unit costs for this work were given in Table 8.7.

Items listed in the general costs sections are taken from the general costs checklist. In this case, no additional form lumber is required, and no additional charges are made for the compressor, vibrators, and curing blankets or for developing power and water because these costs were covered in the estimate for the bridge itself.

We have assumed that no expansion joint material is specified. When this material is required, the labor cost to set it is included in the unit cost to set and strip the barrier railing. In all cases, the costs to set and strip bulkheads at the ends of the walls and at construction or expansion joints are included in the unit costs to

TYPE 3 BARRIER RAILING	MISC.	POSTS	WALK	WALL	I. WALK CY	WALL CY
END POSTS 2(3x1x 3.21/27) 2(3x3.21+4.21x1+0.33x1.54)		29				0.7
WALL A_s = 0.67(2.21+0.33) = 1.70 SF 162 x 1.70/27 162 x 4.75				770		10.2
WALK 2.92 x 0.70 x 168/27 (168+6) 0.7 A_s = 0.75 x 0.5 = 0.375 SF 0.375(28+5.5)/27			122		12.7 0.5	
H EXC. 0.5 CY	1					
FG 28x3 = 84 SF	1					
S19	2	29	122	770	13.2	10.9

	MISC.	MAKE WALK	MAKE WALL	PLY	2X
GENERAL COSTS					
MAKE 82 LF 82x4.75			390		
390+15 = 405 SF x 1.17, 1.5				0.5	0.6
MAKE 15 SF END POST	1				
MAKE (82+3) 0.7 WALK		60			0.2
SAND BLAST 6x1+162x0.7 +140x3 = 540 SF	1				
SACK 162x2.54+29 = 441 SF	1				
TIE HOLES 168/2 = 84,+1+8= 93	1			HAVE	
So	4	60	390	0.5	0.8

MINOR EXPENSES

NAILS 0.024 X 10.5 CWT	40			
HAND TOOLS 0.024 MCY	500			
FORM OIL 92/150 = 6 GAL	0.50	*68		
CHAMFER 3x82 LF	0.13			
DRIVE STUDS 35 EA.	0.30			

Figure 8.41. Takeoff for Type 3 barrier railing.

set and strip the barrier railing walls, and no additional charges are necessary for this work.

The cost to furnish and install the metal railing is a subcontract cost and is normally included in the cost of the barrier railing. In this example we assume a cost of $10.50 per LF of the pay item. Since, due to the concrete end posts, the pay length for barrier railing is greater than the length of metal railing required, it is

Figure 8.42. Takeoff for Type 2 barrier railing.

essential to know exactly what basis is being used by each subcontractor quoting metal railing.

In our estimates we assume the cost of the reinforcing steel to be included in the cost of the barrier railing. This is a subcontract cost and may be quoted on a weight basis, on a linear foot basis, or as a lump sum.

To save time on bid day, the subcontract items may be processed on the estimate

ITEM 48 168 LF TYPE 3-3 BARRIER RAILING

C 16.34 K_{6+2} = 1000 L 13.48 K_{L+F} = 29.58 F 16.10			L	DAYS	M,R
S_0 GENERAL COSTS			$\}$		—
24.1 × 1.02 = 24.75 CY		30			743
MINOR EXP. 168 LF		0.40	$\}$		67
MAKE 390 SF WALL		1.17	456		$\}$
✓ 60 SF WALK		0.25	15		
✓ 15 SF END POST		1.17	18		
SACK 0.441 MSF		500	220		10
FILL 93 TIE HOLES		0.30	28		$\}$
ADD'L. CURE 1 WK. END	½ × 50		25		27
SAND BLAST 540 SF		0.30	162		$\}$
SUBCONTRACT			$\}$		
168 LF METAL RAILING		10.50			1764
24 CWT REINF. STEEL		25.10			602
S_{19} BARRIER RAILING					$\}$
H. EX. 0.5 CY		14	7	162	
FG 84 SF		0.40	34	81	
SS 770 SF WALL		1.60	1232	≈2D	
SS 122 SF WALK		2.50	305		
SS 29 SF POSTS		1.60	46		$\}$
POUR 11.25 CY WALL		43.50	490		
POUR 13.5 CY WALK		26	350		
			3388		3213
INCL. PAYROLL EXP. (1.315)			4455		
Σ = $7668 ($45.64/LF)					

(Left margin at bottom rows: 10.9, 13.2, 24.1)

Figure 8.43. Estimate for Type 3 barrier railing.

ITEM 47 168 LF TYPE 2 BARRIER RAILING			
C16.34 $K_{6+2} = 1000$ L 13.48 $K_{L+F} = 29.58$ F 16.10		L	DAYS	M,R

	S_0 GENERAL COSTS		L	DAYS	M,R
			}		}
	6.9 × 1.02 = 7.25 CY	30			218
	MINOR EXP. 168 LF	0.21	}		35
	MODIFY 390 SF	0.20	78		—
	SACK 277 SF	0.50	138		6
	FILL 8 TIE HOLES	0.30	2		—
	ADD'L. CURE 1 WK. END	½×50	25		—
	SAND BLAST 120 SF	0.30	36		6
	SUBCONTRACT		}		—
	168 LF METAL RAILING	10.50			1764
	15.6 CWT REINF. STEEL	25.10			392
	S_{20} BARRIER RAILING		}		}
	SS 162 LF WALL	6.13	993	162	}
	SS 21 SF POSTS	1.60	34	81	}
6.9	POUR 7.25 CY	43.50	315	≈2 D	}
			1621		2421
	INCL. PAYROLL EXP. (1.315)		2132		
	$\Sigma = \$4553$ $(\$27.10/LF)$				

Figure 8.44. *Estimate for Type 2 barrier railing.*

summary sheet, or assumed costs may be entered on the estimate sheets for the barrier railing and then corrected on the cut sheet as quotations are received.

The time estimates for the barrier railing are made by dividing the linear feet to be installed by our assumed rate of 81 LF per day, We complete the walkway portion of the type 3 railing while constructing the type 2 railing. Under normal circumstances, it is not necessary to allow additional time for installing the metal rail since this work is done concurrently with the final job cleanup or other items of work.

8.1.54 Box Girder Bridges

The box girder bridge is shown in Figure 8.45. Type 1 abutments with the dimensions shown in Figure 8.46 are used. It is assumed that the job conditions and location are the same as those described in connection with the estimate for the retaining walls.

8.1.55 Structure Concrete

It is also assumed that the box girder bridge constitutes the total structure concrete item for the project.

8.1.55.1 THE CONCRETE TAKEOFF. The concrete takeoff is shown in Figures 8.47 to 8.49 and includes the takeoff of the required falsework shown in Figure 8.46. The falsework is based on a preliminary design, and in practice the falsework takeoff would be made from sketches on the design sheets. The preliminary falsework design requires very little time. The toe of the abutment slopes is located, and the shoring towers are spaced so that the footing pads for the end frames nearest to the abutments are, preferably, on the flat surface.

When circumstances require these pads to be located on the toe of the abutment slopes, the toe is removed and later replaced. It is possible to raise the elevation of these end pads by replacing the bottom base frames with inverted extension frames. This procedure permits the end pads to be located on the abutment slopes but requires hand excavation to set the pads; it increases the cost to set and strip the shoring and should be avoided when possible. It is seldom practical to set an entire shoring tower on an abutment slope.

Since the abutments for this bridge are identical to those previously used for the T-girder bridge, the S_{13} total from Figure 8.32 is transferred directly to this takeoff.

To save time and to reduce the likelihood of error, the calculations for the volume of concrete involved in S_{23} and S_{24} do not conform exactly with the true location of the construction joints. The abutment diaphragms are assumed to extend to the top of the top slab, and the height of the girders, bent diaphragm, and intermediate diaphragms are taken as the clear distance between the top and bottom slabs.

The calculations for the contact areas of the required forms are based on the true locations of the construction joints. No deductions are made in the areas of the girder forms at their junction with the intermediate diaphragms, and the effect of

Figure 8.45. Box girder bridge.

237

Figure 8.46. Falsework for box girder bridge.

BOX GIRDER BRIDGE		FG	H 5/10	L. CY
ABUTMENTS & WINGWALLS				
S_{13} (FIG. 4-33)		216	1436	42.8
		MISC	FG	CY
BENT F'ND.				
$14 \times 14 \times 2.5/27$			196	18.2
POUR NEAT ADD'L. CONC. 3.5 CY		1		
$15.3 \times 15.3 \times 2.5/27 - 18.2$				
S_{21}		1	196	18.2
			6'φ H 20?	CY
COLUMN $H = 122.00 - 4.75 - 97.00 = 20.25$				
$6^2 \times π/4 \times 20.25/27$ S_{22}			1	21.2
	MISC.	H 3±	SOF	CY
BOT. SLAB + GIR'S.				
SLAB				
$27.33 \times 175 \times 0.5/27$				88.6
$27.33 \times 175 - π/4 \times 36$			4755	
GIR'S. $L_G = 175 - 6 = 169'$				
$h_c = 4.75 - 1.04 = 3.71'$ $h_f = 3.71 - 0.33 = 3.38'$ INT.				
$5(169 \times 0.667 \times 3.71/27)$				77.4
$2(3.83 \times 169) + 8(3.38 \times 169)$		5864		
FLARES ADD'L.				
$4(3.71 \times 0.25 \times 16/27) + 6(3.71 \times 0.33 \times 16/27)$				6.6
ABUT. DIA.				
$2.5 \times 4.75 \times 32/27$				14.1
$4.75 \times 30 + 4.21 \times 32 + 4.75 \times 5$				
$-27.33 \times 0.5 - 0.67 \times 3.71 \times 5$		275		
1 ADD'L ABUT. DIA.		275		14.1
BENT DIA.				
$6 \times 3.71 \times 27.33/27$				22.5
$2(27.33 \times 3.71) + 2(6 \times 3.83) - 10(6.67 \times 3.71)$		224		

Figure 8.47. Concrete takeoff for box girder bridge.

the girder flares is neglected in determining the contact area of the bent diaphragm forms. The additional thickness of the deck overhang is also neglected in determining the contact areas of the abutment diaphragm forms.

In our case the method used to calculate the concrete volumes increases the true S_{23} volume by 14 CY and decreases the true S_{24} volume by the same amount. Be-

BOX GIRDER BRIDGE (CONT'D.)		MISC	H 3½	SOF	CY
INTERMEDIATE DIA'S.					4.4
8 (6.00 × 3.71 × 0.667/27)					
8 (6.00 × 3.71 × 2)			356		
POUR BULKHEADS					
SLAB 27.33 × 0.5 = 14 SF ⎫					
GIR'S. 5 × 0.67 × 3.38 = 11 SF ⎬		1			
	S_{23}	1	6994	4755	227.7

		H7" DE	OH SOF	LOST SOF	CY
TOP SLAB					
32 × 175 × 0.542/27					112.4
169 × 6.00 × 4				4056	
ADD'L. @ OH (2.33 × 0.71 × 175/27) 2					6.3
175 × 2.33 × 2			816		
175 × 0.583 × 2		204			
FILLETS L = 8 × 169 + 8 × 4 × 6 = 1544'					
1544 × 0.333² × ½ × 1/27					3.2
1544 (0.333√2 − 0.333)				213	
	S_{24}	204	816	4269	121.9

		MISC.	2B+E SS	4×6	4×12 6×12
$\Sigma_{21}^{24} S + S_{13} = 431.8$ CY					
FALSEWORK					
BENT IN 2B+E 2 @ 12.32 CWT			24.6		
4 × 12 4/12				0.19	0.19
6 × 12 4/12				0.29	0.29
4 × 6 28/6				0.34 0.34	
GIR'S.					
IN 2B+E 10 @ 13.28 CWT			132.8		
N 2B+E 20 @ 6.58 CWT			131.6		
END CAPS 4×6 68LF				0.14 0.14	
FLAT PADS (4 SF) 20 EA		1		— 0.3	
6 × 12 100/12				7.20	7.20
4 × 12 20/10				0.80	0.80
16 WF 36 20/26 = 187.7 CWT		1			
TIES 4×6 60 LF				— 0.12	
RODS ADD'L. (HAVE)		1			

Figure 8.48. Concrete takeoff for box girder bridge.

cause of differences in the pour costs, an error of $86 is introduced into the estimate, but this discrepancy is on the low side and is considered insignificant.

The areas of the required slab and girder pour bulkheads are noted in the miscellaneous column of section S_{23} since we plan to make two separate pours for this work.

BOX GIRDER BRIDGE (CONT'D)

	MISC	2B+E	SS	4×6	3. 4×12 6×12

4×6   90-25-3=84.5, ÷1.33=63.5
      (65+1)2 = 132 LINES

```
       14'2        10'        14'2
   | 4' |2.7| 4' |2.7| 4' |2.7| 4' |2.7| 4' |
```

4×6   264/14   132/10

	MISC	2B+E	SS	4×6	6×12
				10.03	10.03
S$_{25}$	3	289.0	19.0	10.9	8.5
				HAVE	HAVE
	MISC	MAKE	HAVE	LOST PLY	PLY	2X

GENERAL COSTS

Item	MISC	MAKE	HAVE	LOST PLY	PLY	2X
S$_{13}$ ABUT'S. MAKE 1 718 × 1.17, 1.75		718			0.84	1.26
S$_{22}$ COL. 1 EA 6'φ×21.25'	1					
S$_{23}$ BOT. SL.+GIR'S. SOF 4755×1.17					5.56	
OS GIR COMPLETE 2×3.83×169 = 1295						
110 LF I.S. GIR (3.38×110)8 = 2974						
2 ABUT DIA 550						
BENT DIA 224						
½ INT. DIA 178						
* 1.17, 1.75 × 5221		5221			6.11	9.14
S$_{24}$ DECK SLAB DE 204×1.17, 1.75		204			0.24	0.36
OH SOF 816×1.17, 4.8					0.95	3.9
LOST SOF 4269×1.17, 1.35				5.0		5.8
RE-USE S$_{13}$ (0.84, 1.26)×2/3					(0.56)	(0.84)
✓ S$_{23}$ 2974+½×550+224+178=3651						
3651×2/3×1.75						(4.3)
✓ × 1.17×10%±				(0.5)		
S$_{25}$ FALSEW'K. I HAVE			19.4	(HAVE 2)	(HAVE 6.0)	
289 CWT 2B+E, 187 CWT 16WF	2					
* TAPER TIES 5221÷8=653	1					
S$_0$	4	6143	19.4	4.5	13.1	15.3

Figure 8.49. *Concrete takeoff for box girder bridge.*

In section S_{25}, the 4 × 6 is listed separately from the 4 × 12 and 6 × 12 stringers since the cost to buy these materials would be different. Upon determining the amounts required, however, we note that we have sufficient material in our yard. In taking off the 4 × 6, we add one additional line of 4 × 6's for each span, and these 4 × 6's will be used under the intermediate diaphragms.

Figure 8.50. *Box girder bridge, Bakersfield, California. Contractor: Tumblin Company.*

The flat pads required at the end caps are noted, and the requisite material is listed. We assume we have these pads in our yard, and no charge is necessary to make them.

As noted in Table 8.10 (see the next section) and Figure 8.17, no additional charges are made for setting, stripping, or hauling the footing pads required for steel shoring or for wooden bent cap falsework. In both cases the pads are set on essentially flat surfaces, no "slope pads" with their attendant hand excavation are involved, and the costs to set and strip the pads are included in the unit costs to set and strip the falsework.

When wooden falsework is used in connection with slab or girder bridges, the cost to set and strip the pads is based on separate unit costs since such pads may be set on flat or sloped surfaces. Wooden falsework may be used on abutment slopes when steel beams are not available, but the cost will be greater than that realized when such falsework is eliminated through the use of steel beams.

In our case, we assume we have the necessary pads with the steel shoring and no additional charges for furnishing them will be required.

In section S_0 the areas from sections S_{13} and S_{21} to S_{24} are converted into MFBM of 2 × 4 and MSF of plywood, and the areas of the form panels to be made are listed in the "make" column.

We plan to make form panels for all the exterior girder faces since these forms remain in place until the top slab is stripped. Since we plan to make two pours, we will make forms for only 110 LF of each interior girder face.

An allowance of 4.8 MFBM of lumber per MSF of deck overhang soffit is used based on the determination made in connection with the T-girder bridge. The allowance of 1.35 MFBM per MSF of lost soffit is based on many past designs.

Allowances are made for reuse of form materials from sections S_{13} and S_{23}. In constructing box girder bridges, the interior girder and diaphragm forms are stripped, but the interior soffit forms remain in place and are lost.

In this instance, we estimate that 4.3 MFBM of material will be available for use in the lost soffit formwork. Although we have more than 4 MSF of plyform available, we consider this too valuable to lose since we have scrap plyform in our yard. Therefore, we allow only about 10 percent of this material for reuse and plan to use the scrap plyform for the remainder.

If scrap plyform were not available, a decision would be required regarding reuse of the $\frac{5}{8}$-inch material from the girders versus the purchase of the cheapest grade of $\frac{3}{8}$-inch interior plywood. In many cases $\frac{3}{8}$-inch plyscord is used in lost soffit formwork.

8.1.55.2 THE STRUCTURE CONCRETE ESTIMATE. The structure concrete estimate is shown in Figures 8.51 and 8.52, and the unit costs for this work are given in Tables 8.5, 8.6, and 8.9–8.11.

The allowance for additional curing depends on the arrangements we can make to have a workman tend to the cure over the weekends. In our case we have assumed that a nearby water supply is available and have allowed $100 to purchase the water and $150 for water hoses and nozzles. On many jobs a rental charge to cover the cost of water trucks, storage tanks, and pumps may be required.

The small allowance for hoppers is made to cover the collection hopper and elephant trunks required to pour the bent column, and we assume we have this equipment in our yard.

The miscellaneous iron item allows 4 hours of welder time to modify the rod bracing for the top flanges of the wide-flange beams. We assume we have rods used on past work in our yard.

The 41.5 MFBM of material noted in the cleanup item represents the sum of the MFBM and MSF of material listed to be furnished less the 10.8 MFBM of material in the lost soffit formwork. For this calculation 1 MSF is assumed to equal 1 MFBM.

In section S_{21} we increase the unit cost to set and strip the bent foundation by 10 percent because our foundation is somewhat larger than average.

The unit costs shown for setting and stripping the girder forms include the costs to set and strip formed drains through the bottom slab. One or two 4-inch-square openings are normally specified for each box girder cell.

When utility openings through the diaphragms are specified, the unit cost to set and strip the necessary forms will be about 5 times the unit cost used to set and strip the girder forms.

When hinges are specified, the unit cost for setting and stripping the hinges will be the same as the unit cost for setting and stripping the girders. An additional allowance must be made to furnish and place neoprene pads, hinge armor, or hinge tie rods when required.

In section S_{23}, no charge is made for setting and stripping the plyform under the bottom slab. This material is located directly on the falsework, and the cost to set and strip it is included, as always, in the unit costs to set and strip the falsework.

The time required to complete the various operations is determined in the "days" column. A two-carpenter crew is used on the bent foundations, four-carpenter crews are used on the abutment and column work, and six-carpenter crews are used for the bottom slab plus girders, top slab, and falsework.

ITEM 37 430 CY STRUCTURE CONC.			
C 16.34 K$_{6+2}$ 1000 L 13.48 F 16.10 K$_{cr}$ 75	L	DAYS	M, R
S$_0$ GENERAL COSTS			
CONC. 431.8×1.02 = 441 CY 30			13230
15.3 MFBM 2X 300			4590
19.4 ″ HAVE 120			2330
7.1 MSF PLY 500			3550
6.0 ″ ″ HAVE 170			1020
4.5 ″ LOST PLY HAVE 50			225
289 CWT 2B+E 2 MOS. 578			1160
187 CWT WF I USE. 1.50			280
COL. FORM 6'ø x 24' I WK 450			450
6.5 C TIES 2 MOS. 40/c			520
MINOR EXP. 0.432 MCY 2370			1030
MAKE 6143 SF 0.26	1600		—
ADD'L. FIN. 0.432 MCY 3630	1570		60
ADD'L. CURE 5 WKS 90	450		—
SAND BLAST 0.432 MCY 872	375		65
COMPRESSOR 2 MOS 400			800
VIBRATORS ″ ″ 200			400
DECK FIN. MACH. 5760 SF 0.05			290
GENERATOR 2 MOS. 200			400
DEVELOP WATER 100+150			250
HOPPERS			20
CURE RUGS			130
MISC. IRON 4 HR WELDER 30			120
BOX GIR. DOBIES 0.228 MCY 700	160		10
SWEEP 5.76 MSF 13	75		—
CLEAN UP 41.5 MFBM 43 12	1780		500
23.8 T 34 7	810		165
S$_{13}$ ABUT. + WW			
FG 216 SF 0.40	85	2885	
SS 1436 SF H 5/10 1.95	2800	1000×2/3 =5 D	
428 POUR 44 CY 5.25	230		
S$_{21}$ BENT F'ND.			
FG 196 SF 0.28	55	160	
3.5 CY ADD'L. CONC. I 30	5	1000×1/3 =1 D	105
SS I F'ND. 110% x93	100		
POUR 19 CY 4.45	85		

Figure 8.51. *Estimate for box girder bridge.*

One additional working day is added to the bottom-slab-plus-girder time to allow for the first pour. Nine working days are added to the time required to set and strip the top slab, 1 working day for the slab pour, and 8 working days to comply with a specified 10-calendar-day delay before starting the stripping operation.

The estimated time to complete the item is 45 working days, as shown in Figure 8.52.

ITEM 37 430 CY STR. CONC. (CONT'D)

				L	DAYS	M,R
	S_{22} BENT COL.					
	SS 1 COL.	360	150	360	$\dfrac{360}{1000\times 2/3}$ =1D	150
	CRANE ON-OFF ADD'L 2 HR		75	—		150
21.2	POUR 22	8.30	10.50	185		230
	S_{23} BOT. SLAB+GIR'S.					
	SS SLAB BULKH'D. 14 SF		5.30	75	8130	
	SS GIR BULKH'DS. 11 SF		1.15	15	1000	
	SS 6994 SF H34½		1.15	8040	=9,+1	
227.7	POUR 232 CY	7.75	5.80	1800	10D	1350
	S_{24} TOP SLAB					
	SS 204 SF DE		4.25	865	6615	
	SS 5085 SF (LOST+OH) SOF.		1.13	5750	1000 =7,+1+8	
121.9	POUR 124 CY	11.20	7.05	1390	16D	875
431.8	**S_{25} FALSEWORK**					
	SS 289 CWT 2B+E	8.90	0.60	2570	7520	175
	SS 20 16WF 36-26'	40	29	800	1000	580
	SS 20 F PADS		7	140	=8D	—
	SS 19.0 MFBM	211	54	4010		1030
				36180		36240
	INCL. PAYROLL EXP. (1.315)			47575		

Σ = $ 83,815 ($194/CY)

TIME EST.	CONSEC.	CONC'T.
ON-OFF	3	3
ABUT'S.	5	3
BENT F'ND + COL	2	2
BOT. SL. + GIR'S.	10	10
TOP SLAB	16	16
FALSEW'K	8	5
REINF.		
ABUT'S	1	
COL	1	6
BOT SL. + GIR'S	5	
TOP SL.	3	
	54 D	45 D

Figure 8.52. Estimate for box girder bridge.

8.1.56 Prestressed Box Girder Bridges

The procedures and unit costs necessary to estimate the cost of prestressed box girder bridges are identical to those described in the preceding except that some additional factors must be considered.

Although the specifications must be checked in each case, there is normally a

TABLE 8.9 Unit Costs for Box Girder Bridges

Activity	Labor Unit Cost ($)	Index	Rental Unit Cost ($)	Index	Unit of Measure
Additional finishing	3630	$K_{L+F} = 29.58$	—		MCY
Sandblast	872	$K_L = 13.48$	—		MCY
Additional cure	775	$K_L = 13.48$	a		MCY
Make girder forms	2770	$K_{6+2} = 1000$	—		MCY
	0.26	$K_{6+2} = 1000$	—		SF
SS girder forms and pour bulkheads	1.15	$K_{6+2} = 1000$	0.06	$K_{Cr} = 75$	SF
SS slab pour bulkheads[b]	5.30	$K_{6+2} = 1000$	—		SF
Pour bottom slab plus girders	7.75	$K_{L+F} = 29.58$	5.80	$K_{Cr} = 75$	CY
SS lost soffit and OH soffit[c]	1.13	$K_{6+2} = 1000$	—		SF
SS deck edge H6″–H7″	4.25	$K_{6+2} = 1000$	—		SF
Pour deck slab	11.20	$K_{L+F} = 29.58$	7.05	$K_{Cr} = 75$	CY

[a]Rental cost will depend on job conditions. See text.
[b]Increase unit cost 30 percent when bulkhead is skewed with respect to the transverse reinforcing steel.
[c]Increase unit cost 25 percent on skewed bridges when skew is greater than 20 degrees.

Note: The equipment costs for pouring assume normal access for the crane, and the concrete can be deposited directly into the forms. When pours are large or average in size, the unit costs for the crane include 1 hour of travel time. On small pours, travel time will be additional.

single, separate job item for prestressing cast-in-place concrete, and this item will cover all of the prestressing required on the project.

Prestressing is done by subcontractors, and different subcontractors often quote different prestressing systems.

The subcontract price usually includes the cost to furnish design calculations relating to the prestressing system being used as well as the complete cost to furnish and install the necessary ducts, tendons, and vents, including the costs to tension and grout the tendons.

TABLE 8.10 Unit Costs to Set and Strip Bridge Falsework Consisting of Timber on Shore ''X''[a] Towers

Number of Shoring Units (in height)	Steel Shoring Towers ($/CWT) Labor $K_{6+2} = 1000$	Rental $K_{Cr} = 75$	Top Lumber[b] ($/MFBM) Labor $K_{6+2} = 1000$	Rental $K_{Cr} = 75$
1	10.10	—	148	—
3	8.90	0.60	211	54
5	11.50[c]	0.60	300	54

[a]The trademark Shore ''X'' is registered by Waco Scaffold & Shoring Co.
[b]Costs are based on using 4 × 12 or 6 × 12 stringers with a maximum length of 12 feet. Transverse beams range from 4 × 4 to 8 × 8 inches in cross section.
[c]Towers four or more frames in height require 2 × 6 cross bracing. Allow $745 per MFBM ($K_{6+2} = 1000$) for setting and stripping the 2 × 6 bracing.

Note: The unit costs shown include the costs to set and strip the footing pads for the towers and 1 hour of travel time for the crane. Increase the unit costs for labor by 65 percent when small amounts of falsework are used to form isolated bent caps. On such work, the on-and-off charges for the crane and the shoring must be added.

TABLE 8.11 Miscellaneous Falsework Costs

	Labor		Rental		
Activity	Unit Cost ($)	Index	Unit Cost ($)	Index	Unit of Measure
SS wide flange beams[a]					
16 WF 36–24'	40	$K_{6+2} = 1000$	29	$K_{Cr} = 75$	Each
24 WF 76–25'	59	$K_{6+2} = 1000$	55	$K_{Cr} = 75$	Each
Footing pads					
Make	115	$K_{6+2} = 1000$	—		MFBM
SS flat pads 2' × 2'[b]	7	$K_{6+2} = 1000$	—		Each
Hand excavation for pads on abutment slopes	21	$K_L = 13.48$	—		CY
Hand backfill for pads on abutment slopes	24	$K_L = 13.48$	—		CY

[a]Labor costs include the cost to set and strip lateral bracing for top flanges of beams. Rental costs assume easy access for crane.
[b]Pads set on flat surface.

Note: When 20 or more beams will be set or stripped at one time, the unit costs for the crane will include 1 hour of travel time. Unit costs to set and strip wood falsework are shown in Figure 8.17.

8.1.57 Work Area and Water Supply

Prestressing subcontractors require sufficient work areas to enable them to stretch out and measure the ducts and tendons. On a typical bridge job this does not entail any additional cost to the general contractor since the bridge approach roadways are available for this purpose.

In addition, the general contractor may be required to furnish the water required for the grouting operation.

Figure 8.53. *Box girder bridges, Kern County, California. Contractor: Tumblin Company.*

8.1.58 Additional Reinforcing Steel

The specifications may require additional longitudinal reinforcing steel to be draped in the girder stems to facilitate the securing of the ducts. The cost of such reinforcing steel is frequently included in the prestressing item but is usually excluded by the prestressing subcontractor. In such cases the cost to furnish and place this reinforcing represents an additional cost to the general contractor.

8.1.59 Vents

Vent pipes may be required. The vents are made of $\frac{1}{2}$-inch-diameter iron or plastic pipe and extend from the ducts upward through the top slab. After the tendons have been grouted, the vent pipes are cut off 1 inch below the top surface of the slab, and the deck is patched.

The general contractor normally blocks out the top slab to a depth of 1 inch prior to the pour and subsequently cuts the vent pipes and patches the blocked-out portion of the deck. Based on $K_{6+2} = 1000$, the cost for this work will run about $60 per girder per set of vents.

8.1.60 Girder Bulkheads

On prestressed box girder bridges, the ends of the girders normally extend a minimum of 6 inches beyond the outer face of the abutment diaphragms. On skewed bridges, the girders are extended as required to enable the end bulkhead to be set perpendicular to the girder centerline. In addition, a section through the end bulkhead itself will be stepped or parabolic in shape. This is necessary to obtain bearing surfaces for the tendon-locking devices, which will be perpendicular to the draped tendons. The details for the end bulkheads vary with the prestressing system being used, and the exact dimensions are furnished by the prestressing subcontractor.

The unit cost to make, set, and strip the formwork for the extended girder sides and ends runs about six times the unit cost used to set and strip the typical girder forms, and this cost will include the cost of laying out and drilling the holes required in the end bulkheads for the ducts.

After the bridge has been prestressed, the plans may require the locked tendon ends to be encased in concrete. The unit costs shown in Table 8.6 for curtain walls may be used to arrive at an estimated cost for this work.

8.1.61 Falsework during the Tensioning Operation

The box girder bridge will decrease in length and will rise, or hog up, at midspans as the girders are tensioned. The maximum horizontal movement from the reduction in length will occur at each end of the section being tensioned, or, typically, at the abutments.

As the bridge becomes shorter, the X braces in those shoring towers nearest the abutments will start to bow and must be removed as necessary. Since the load on the towers is being reduced as the girders are being tensioned, there is no danger involved, but the operation should be supervised by a competent employee. The cost of the workman used to remove the X braces represents an additional cost to the general contractor.

8.1.62 Precast Concrete Girders

A typical precast concrete girder is shown in Figure 8.54. Such girders are frequently used in constructing highway bridges. Their heights normally vary from 2.75 to 5.5 feet, and their lengths range from 50 to 120 feet.

When precast concrete girders are required, separate job items are set up for furnishing and erecting them. Thus, there may be one item for furnishing 15 girders 55 to 60 feet in length, a second item for furnishing 12 girders 60 to 65 feet in length, and a third item for erecting 27 girders 55 to 65 feet in length.

Both the furnishing and the erecting items are bid on a price-per-girder basis, and the price for furnishing the girders normally includes the costs for prestressing them.

Except for projects in remote locations, it is usually cheaper to have precast concrete girders furnished by precast concrete manufacturers. Savings are realized through the use of long casting beds which are set up for pretensioning, use of

Figure 8.54. Precast concrete girders.

metal forms, availability of overhead gantry cranes, and use of sophisticated concrete mixing, placing, and vibrating equipment. In addition, the girders are usually steam cured.

Manufactured girders should always be erected by the manufacturer since it is never wise to split the responsibility for a product unnecessarily.

Manufactured girders up to about 85 feet long are delivered to the job in one piece by trucks. Delivery is scheduled to fit the erection sequence, and the girders are taken from the trucks and placed in position in one operation.

Precast concrete I-girders over 85 feet long require special handling during the trucking operation to prevent buckling. Because of the added costs and risks associated with hauling the longer girders, manufacturers often fabricate them in two pieces, which are joined and post tensioned at the erection site.

The erection is considered to be completed when the girder has been set in its plan location and the slings have been removed. It is the general contractor's responsibility to tie off and brace the positioned girders prior to the removal of the slings. Unit costs for this work are shown in Table 8.12.

Care and good judgment must be exercised during the girder bracing operation since until the girder diaphragms are poured, the girders are somewhat unstable and are vulnerable to earthquake and, in the case of longer girders, to buckling.

The manufacturer's quoted price per girder should include the cost of furnishing holes through the exterior girder webs as required for attaching the deck overhang brackets as well as the cost to furnish the inserts and holes necessary to tie the

TABLE 8.12   Unit Costs for Precast Prestressed Concrete I-Girders

	Labor		Rental		
Activity	Unit Cost ($)	Index	Unit Cost ($)	Index	Unit of Measure
FG, SS, P & F casting bed slabs	1.34	$K_{L+F} = 29.58$	—		SF
Make side forms H < 4'	4.00	$K_{6+2} = 1000$	—		LF[a]
H4'–H5.5'	1.00	$K_{6+2} = 1000$	—		SF
Make end bulkheads	5.65	$K_{6+2} = 1000$	—		SF
SS side forms H2.75'	0.70	$K_{6+2} = 1000$	—		SF
H4'–H5.5'	1.05	$K_{6+2} = 1000$	—		SF
SS end bulkheads	4.75	$K_{6+2} = 1000$	—		SF
Form holes or set inserts	2.90	$K_{6+2} = 1000$	—		EA
Set beginning plates or cable loops	1.50	$K_{6+2} = 1000$	—		EA
Pour girders	10.50	$K_{L+F} = 29.58$	—		CY
Erect girders, L50'–L85'[b]					
Crane rental	—		1.58	$K_{Cr} = 75$	LF
Truck rental	—		0.63	$K_{Tr} = 30$	LF
Rigger labor	0.72	$K_R = 17.00$	—		LF
Tie off girders	0.66	$K_{6+2} = 1000$	—		LF

[a]Per LF of form panels required for each side.
[b]See text for assumed conditions. On-and-off charges for the equipment and the cost to rig cranes, when required, are additional. If girders are cast at the erection site, eliminate the truck cost and reduce crane rigger costs by 50 percent.
Note: Unit costs to set neoprene pads and celotex are shown in Table 8.13.

Figure 8.55. *Precast I-girder bridge, Beardsley Canal. Contractor: Tumblin Company.*

poured-in-place diaphragms to the girders. (See Figure 8.57.) In addition, the manufacturer's price should include the cost of sandblasting the tops of the girders.

In all cases, the general contractor must contact each girder manufacturer before the bid date to determine exactly what will be included in each of their quotations.

8.1.63   Job-fabricated Precast Girders

Precast girders, when fabricated on the job, are normally post tensioned. In such cases the prestressing work is usually performed by a prestressing subcontractor

Figure 8.56. *Precast I-girder bridge, U.S. Highway 99. Contractor Tumblin Company.*

whose price will include the costs to furnish design calculations relating to the prestressing system being used, to furnish and install the necessary ducts and tendons, and to tension and grout the tendons.

The deformed bar reinforcing required for job-fabricated girders is furnished and placed by the subcontractor used on the bridge work, but the cost is normally included in the cost of furnishing the girders.

Although the specifications must be checked in each case, the cost to furnish embedded miscellaneous iron such as bearing plates is usually included in the costs to furnish the girders.

8.1.63.1 THE CONCRETE TAKEOFF. The unit costs to fabricate precast girders at the job site are shown in Table 8.12, and the determination of the quantities relating to the activities listed constitutes the major portion of the concrete takeoff. The following items must also be considered in estimating the cost of precast girders.

8.1.63.2 THE CASTING YARD. The casting yard site must be obtained, and it should be located as close to the point of erection as possible. The number of casting beds required depends on the girder erection schedule, and in many cases some or all of the casting beds may be used more than once. Two lines of casting beds may be set 7 feet center to center with 20-foot-wide roadways at each side, and this pattern is repeated as necessary.

The casting beds should be set low enough to permit the girders to be poured directly from the transit-mix trucks.

The costs to grade the site as well as to dispose of the concrete casting beds on completion must be included in the cost to furnish the girders. The costs for such work are discussed in the chapter 4 on earthwork.

The lengths of the girders will decrease during the tensioning operation, the midportions of the girders will rise up off the casting beds, and the girders will be bearing only at their extreme ends. A short piece of plywood is embedded in the top surface of the casting bed at the girder ends to prevent damage. The hogged-up girders may require temporary ties to ensure stability. The cost of this work is nominal and is included in the costs of the other activities.

8.1.63.3 GIRDER ERECTION. The unit costs shown for erecting girders are based on using two cranes for each girder. It is assumed that the bridge is low enough to permit the cranes to erect the girders while setting below the bridge deck and that normal access is available to this area for the trucks and cranes.

It is assumed further that the casting yard is approximately 400 feet from the erection site and the job conditions are such that the cranes cannot walk the girders to the point of erection. Two trucks are used, and the girders are loaded onto the trucks and subsequently erected with the same cranes.

In all crane work, it is imperative that the cranes selected have ample lifting capacities to handle the existing loads when their booms are lowered, as required by the job conditions. As a rule of thumb, the cranes selected should furnish twice the lifting capacity theoretically required.

On bridges having large skews, it may be impossible to erect girders with only two cranes. In such cases it may be necessary to transfer one girder end to a third crane located behind the bridge abutment.

When erection conditions are difficult, a plan view of the proposed crane is drawn on tracing paper to the proper scale, and this model is shifted over the plan view of the girders and the abutment fill to determine the required horizontal projection of the boom.

The hauling and the erection of I-girders over 100 feet long present special problems. Even though great care has been exercised in arranging the ducts, such girders may bow into flat C or S shapes when tensioned. In such cases lateral hog rod bracing may be required near the bottom flange before moving the girders from the casting beds, and this bracing must remain in place until the girders have been set and tied off. The costs to furnish this bracing and the additional time required to place and remove it increase the costs to furnish and erect long girders.

Due to the difficulties involved, the costs of cranes, trucks, and riggers required to erect 120-foot girders may run from three to three and a half times the costs shown in Table 8.12 for the shorter girders, and the cost to tie off and brace the 120-foot girders will run about twice those shown.

Flagmen and/or detours may be required during the erection operation. Flagmen always constitute an added expense, but the detour work may be paid for at the bid unit prices for the work involved.

8.1.63.4 GENERAL COSTS. The general costs section includes those items from the checklist that have not been covered in the estimate for the structure concrete item.

The amount of forms to be made depends on the job schedule and on the number of girders required. Two and a half sets of forms will enable two girders 60 to 80 feet long to be completed per day, or one girder 120 feet long to be completed on alternate days.

When forms are reused twelve or more times, it may be necessary to grind the girder sides lightly for uniform appearance. In all cases the exterior face of the exterior girders is sacked together with the deck edge after the bridge deck has been completed.

The girders are easily cured with soaker hoses and polyethylene film.

The top surfaces of the girders must be sandblasted, and when large areas are involved, this work is done by sandblasting subcontractors.

Special vibrators are required for concrete girders because of the reduced clearances at the ducts. Two vibrators with $\frac{3}{4}$-inch-diameter spuds will be required for use in the worst places. Heavy vibration is necessary since high strengths and low slumps are involved.

Care must be taken to ensure that the forms do not rise during the pour. When such rise does occur, the bottom chamfers will require additional finishing, which may cost as much as $0.85 per LF of girder based on $K_F = 16.10$.

A dependable bond breaker is required on the top surface of the casting bed. During rainy weather, it may be necessary to lay plywood strips on the casting bed to prevent adhesion.

Figure 8.57. Precast concrete girder bridge.

254

Figure 8.58. *Soffit forms for precast concrete girder bridge.*

A loop of wire rope is embedded at each end of the girder for lifting. These loops are located on the top surface and are on the girder centerline. The cost to furnish the wire rope used is included in the general costs section.

8.1.64 Precast Concrete Girder Bridge Decks

A precast concrete girder bridge is shown in Figure 8.57, and the deck soffit forms are shown in Figure 8.58. The concrete takeoff for the bridge deck is shown in Figures 8.59 and 8.60 and includes the takeoff for the soffit forms. Unit costs for the bridge deck work are given in Table 8.13.

The following discussion is limited to the costs of constructing precast concrete girder bridge decks since takeoffs and estimates for abutments, bents, and complete bridges have been illustrated in previous sections. The furnishing and erecting of the girders themselves is a separate job item, as was discussed in the preceding section.

8.1.65 Unit Costs

The unit cost for pouring the diaphragms is based on doing this work with concrete buggies running on the deck soffit forms.

The cost to make up the deck overhang brackets is carried separately since these brackets may be reused when several similar bridges are being built.

The unit cost to set and strip the soffit forms is applied to the gross area of the deck slab, and no deductions are made for the plan areas of the girders or diaphragms. The unit cost to set and strip the soffit forms includes the costs to set and strip all the formwork shown in Figure 8.58, including the deck overhang brackets, hangers, rods, plywood, and chamfer strips.

PRECAST CONC. GIRDER BRIDGE	MISC.	GROSS SOF	H4	I. CY
<u>DECK SLAB</u>				
ABUT. DIAPHRAGMS				
2 × 4.17 × 28.25/27 − 5 × 0.75 × 4 × 1.58/27				7.8
28.25 × 4.71 + 4 × 4 + 4 × 5.09 × 4.17			234	
1 ADD'L.			234	7.8
SET 64 LF EXP. JT. ARMOR	1			
= 6 PCS. @ 10.7' 8.84 CWT. EA }				
FURN. & SET 40 - ¾"Φ × 16" MB	1			
BENT DIAPHRAGM				
3 × 4.17 × 28.25/27 − 10 × 0.75 × 4 × 1.58/27				11.3
8 × 5.09 × 4.17 + 2 × 3 × 4			194	
INT. DIAPHRAGMS				
A = 5.09 × 2.92 + 2.25 × 0.5 × 2 = 17.1 SF				
17.1 × 0.667 × 16/27				6.8
17.1 × 2 × 16			547	
SS 16 SOF'S. @ 6.09 LF EA.	1			
DECK				
32 × 0.542 × 83.58 × 2/27		5349		107.4
ADD'L @ GIR'S.				
0.167 × 1.58 × 80.08 × 10/27				7.8
ADD'L. @ OH (7 + 11½) ½ = 9.25" = 0.77'				
(0.77 − 0.54) 1.88 × 83.58 × 2 × 2/27				5.4
SS DE 83.58 × 2 × 0.585 × 2 = 195 SF	1			
* DIA'S. 33.7 CY $26	4	5349	1209	154.3*
DECK 120.6 CY				
NEOPRENE PADS 12 × 1½ × 22 20 EA				
1½" CELOTEX (2 × 29)2 + 3 × 29 = 203				
FURNISH 7 × 32 = 224 SF				

Figure 8.59. *Concrete takeoff for precast concrete girder bridge deck.*

PREC'ST CONC. GIR. BRIDGE (CONT'D)	MISC.	PLY.	4×6 4×4	2. 2×8 2×6
SOFFIT FORMS				
X BMS. $^{78}/_6$ =13, +1 = 14 SETS EA. SPAN				
14×2×4 = 112 SETS 2×8 $^{224}/_6$				1.8
INT. JOISTS 14×2×5×4 = 560				
$^{560}/_5$ = 112 2×4$^{112}/_9$, 2×6 $^{560}/_6$				4.0
HANGERS 14×2×5 = 140 EA	1			
2×6 .SPA. 140 × $^8/_5$ = 224 PCS				
= 0.2 MFBM HAVE	1			
SHE-BOLTS 140 SETS	1			
½"ø RODS $^{56}/_8$" $^{224}/_6$" ⎫	1			
✓ ✓ NUTS 280 EA. ⎭				
OH BRACKETS				
$^{78}/_6$ = 13 EA SPAN, + 1 ⓪ BENT				
= 27 EA SIDE = 54 TOT.				
4×4 $^{54}/_8$			0.6	
2×4 $^{54}/_3$				0.1
MAKE 0.7 MFBM OH BR'K'S.	1			
3×3×¼ ∟ 54 PCS 4"LG-3 HOLES ⎫				
½"ø RODS $^{54}/_{40}$"				
✓ ✓ NUTS 108 EA ⎬	1			
✓ ✓ CUT WASH. 54 EA				
✓ ✓ INSERTS 2 EA ⎭				
OH JOISTS 28 SPA. ⓪ 6' EA SIDE				
2×6 28×2×3×6 = 1008 LF				1.0
4×6 28×2×1×6 = 336 LF			0.7	
2×4 BRIDG'G. 336 LF				0.2
BLOCKING NC HAVE				
PLYFORM				
82.08×2×5.09×4 + 83.58×2×1.88×2				
= 3971 SF, × 1.17		4650		
INT. DIA. SOFFITS 2×6 $^{32}/_6$				0.2
S_{27}	6	4650	1.3	7.3

Figure 8.60. *Concrete takeoff for precast concrete girder bridge deck (continued).*

TABLE 8.13 Unit Costs for Precast Concrete Girder Bridge Decks

Activity	Labor		Rental		Unit of Measure
	Unit Cost ($)	Index	Unit Cost ($)	Index	
SS intermediate diaphragm soffits	8.00	$K_{6+2} = 1000$	—		LF
Make diaphragm forms	0.26	$K_{6+2} = 1000$	—		SF
SS diaphragm forms	2.12	$K_{6+2} = 1000$	—		SF
Pour diaphragms	7.60	$K_L = 13.48$	—		CY
Make deck overhand brackets	450	$K_{6+2} = 1000$	—		MFBM
SS gross deck soffit forms	1.51	$K_{6+2} = 1000$	—		SF[a]
SS deck edge H6"–H7"	4.25	$K_{6+2} = 1000$	—		SF
Pour deck slab	11.20	$K_{L+F} = 29.58$	7.05	$K_{Cr} = 75$	CY
Set celotex and neoprene pads	0.50	$K_{6+2} = 1000$	—		SF
Set expansion joint armor	4.00	$K_{6+2} = 1000$	4.00	$K_{Cr} = 75$	LF[b]

[a]Per SF of gross soffit area. Gross soffit area $= WL$, where W = overall width of deck slab and L = length of deck slab.
[b]Per LF of armored joint.

Note: The equipment cost for pouring assumes normal access for the crane, and the concrete can be deposited directly into the forms. When pours are large or average in size, the unit costs for the crane include 1 hour of travel time. On small pours, travel time will be additional.

8.1.66 The Concrete Takeoff

The takeoff for the general costs section is not illustrated since it depends on the total amount of structure concrete involved.

Figure 8.57 shows the deck soffit to be 2 inches above the top of the girders. This dimension varies in practice, and an approximate average value must be calculated. The calculation is based on the given elevations for the girder bearings at the abutments and at the bent cap, the given centerline profile, and the given camber diagrams for each span. The camber diagrams take into consideration the estimated amount of camber induced in the girders when they were tensioned.

As a general rule, the primary activities are listed as column headings on the concrete takeoff. Using $8\frac{1}{2} \times 11$ quadrille pads having five squares per inch, as many as ten columns may be used in practice. Due to lack of space in Figure 8.60, however, the intermediate diaphragm soffits and the deck edge are carried in the miscellaneous column.

In this example, no difficulty results from this manner of listing these two items since only one entry is required for each of them. On projects involving more than one bridge, however, the totals for those sections of the individual bridge takeoffs which pertain to work that is common to more than one structure are combined, and the combined takeoff is used to set up the estimate sheets. In such cases, it is important that those primary activities common to more than one structure be listed as column headings on the takeoffs for each structure so as to facilitate the preparation of the combined takeoff.

The neoprene pads as well as the area of the celotex required are noted at the bottom of sheet 1 of the concrete takeoff. In practice, these materials would be included in the sections of the takeoff pertaining to the abutments and the bent cap.

The area of celotex noted to be furnished is based on buying the material in 4 × 8-foot sheets. The unit cost shown in Table 8.13 for placing the celotex should be applied to the total area to be furnished.

8.1.66.1 THE SOFFIT FORMWORK. The soffit formwork is based on a preliminary design, and the required materials would be taken off from sketches on the design sheets.

The rods used with the girder hangers are made up using half a she-bolt together with a $\frac{1}{2}$-inch-diameter tie rod. The coarse Acme threads on the she-bolts permit a quick preliminary grade adjustment to be made from the bottom, and the precise adjustment is made with the $\frac{1}{2}$-inch nut at the top of the hanger. On precast girder bridges, excessive lateral deflection of those interior longitudinal joists located at the sides of the girders must be prevented. In our case, these joists are held firmly against the girder sides with 2 × 4 bridging at 3-foot centers. If the fluid pressure of the concrete causes undue lateral deflection of these joists, leakage will occur, and the construction joint at the tops of the girders will require a considerable amount of additional but needless finishing labor.

8.1.66.2 EXPANSION JOINT ARMOR. The cost to furnish expansion joint armor is normally included in a separate job item for miscellaneous iron. The cost to install armor is included in the estimate section covering the bridge deck.

Expansion joint armor (Figure 8.57) is furnished in two pieces temporarily bolted together for erection. One piece consists of the two plates, and the other consists of the angle and the 1-inch-square bar. Open holes $\frac{3}{8}$, inch in diameter and 18 inches on centers should be provided in both pieces to enable the armor to be firmly secured before placing the concrete.

8.1.66.3 ADDITIONAL FINISHING. The outer face of the exterior girders and the deck edge must be sacked after the slab and barrier railing have been completed. In addition, the tie rod holes in the exterior girders and in the slab soffit must be filled.

8.1.66.4 SANDBLASTING. The construction joint at the top of the girders must be sandblasted. Manufactured girders are usually sandblasted at the factory, but this must be verified in each case.

8.1.67 Plate Girder Bridge Decks

Plate girder bridge decks together with typical soffit formwork are shown in Figures 8.61 and 8.62. The applicable unit costs are given in Table 8.14.

The plate girders themselves are covered by a separate job item for structural steel, which is normally bid on a price-per-pound basis.

Plate girders are furnished and erected by structural steel subcontractors. Anchor bolts, when required, are furnished by the steel subcontractor but are set and dry packed by the general contractor. Steel bearing assemblies are furnished and set by the structural steel subcontractor at the time the girders are erected. Neoprene bear-

Figure 8.61. *Plate girder bridge deck.*

ing pads are normally furnished and set by the steel subcontractor, but this must be confirmed in each case.

The cleaning and painting of plate girders is covered by a separate job item which is usually bid as a lump sum. In most instances, the girders are cleaned and primed in the shop by the structural steel subcontractor, who quotes a separate lump sum price for this portion of the painting item.

The field painting is always done by painting subcontractors. Their quoted prices will be generally proportional to the pounds of structural steel involved.

Figure 8.62. *Plate girder bridge deck (continued).*

TABLE 8.14   Unit Costs for Plate Girder Bridge Decks

	Labor		Rental		
Activity[a]	Unit Cost ($)	Index	Unit Cost ($)	Index	Unit of Measure
Make deck overhang brackets	450	$K_{6+2} = 1000$	—		MFBM
SS soffit forms S3′–S12′[a]					
Gross OH soffit	2.72	$K_{6+2} = 1000$	—		SF
Gross interior soffit	1.16	$K_{6+2} = 1000$	—		SF
SS soffit forms S17′[a]					
Gross OH soffit	3.85	$K_{6+2} = 1000$	—		SF
Gross interior soffit	1.64	$K_{6+2} = 1000$	—		SF
SS beam sides	3.60	$K_{6+2} = 1000$	—		SF
SS deck edge H6″ = H7″	4.25	$K_{6+2} = 1000$	—		SF
Pour deck slab[b]	11.20	$K_{L+F} = 29.58$	7.05	$K_{Cr} = 75$	CY

[a]When S (girder spacing) varies from 3 to 12 feet, the minimum unit cost to set and strip the gross soffit area will be about $1.30 per SF. Average unit costs to set and strip gross soffit areas are:

S3′–S9′	$1.32 per SF
S11′–S12′	$1.44 per SF
S17′	$2.10 per SF

The preceding unit costs are indexed to $K_{6+2} = 1000$.

[b]The equipment cost assumes normal access for the crane, and the concrete can be deposited directly into the forms. When pours are large or average in size, the unit costs include 1 hour of travel time. On small pours, travel time will be additional.

Note: Gross OH soffit area is the sum of the two OH dimensions (Figure 8.6) multiplied by the length of the deck slab. Gross interior soffit area is the width, center to center, of the exterior girders multiplied by the length of the deck slab. Gross soffit area is gross OH soffit area plus gross interior soffit area, or total area of bridge deck.

The primed steel girders will be stained at the time the deck is poured and must be cleaned before the field painting starts. The cost of doing this work has been discussed previously in Section 8.1.1.

8.1.68   Unit Costs

The unit costs for setting and stripping deck soffit forms are based on gross soffit areas defined in the footnote to Table 8.14. No deductions in area are made for the areas occupied by the girder flanges. The unit costs given include the cost to set and strip all the formwork shown, including the plyform and chamfer strips.

The unit costs shown for setting and stripping soffit forms are based on gross soffit areas ranging from 5000 to 16,000 SF. For gross soffit areas of 1000 SF the unit costs for setting and stripping should be increased about 30 percent and the cost to pour the slab should be increased about 50 percent.

The unit cost for setting and stripping beam sides is applied to the vertical contact areas of deck beams such as those occurring at expansion joints.

Separate unit costs are shown for setting and stripping gross overhang soffit forms and gross interior soffit forms. In the field, these operations are performed

Figure 8.63. *Plate girder bridge, Kern River. Contractor: Tumblin Company.*

concurrently by one crew, and it is virtually impossible to ascertain true unit costs for each operation. For the same reason, the unit costs previously discussed for setting and stripping falsework include the costs for setting and stripping that ply-form immediately on the falsework.

Viewed algebraically, it would seem that separate unit costs could easily be derived by the use of simultaneous equations applied to the average unit cost data from two different jobs or decks.

In practice, such procedures may result in derived unit costs which, although mathematically correct, are grossly different from the true unit costs. The reason for such discrepancies is that the true unit costs being realized for the separate operations are never precisely the same on different jobs, nor even from day to day

Figure 8.64. *Plate girder bridge, Olive Drive Overcrossing. Contractor: Tumblin Company.*

Figure 8.65. Plate girder bridge, Haunched Girders. Contractor: Tumblin Company.

on a particular job, since unit costs are influenced by many varying factors, as discussed in Chapter 3. An example clearly illustrates the problem.

Assume that two similar plate girder bridge decks have been constructed and that the dimensions and average unit costs realized were as follows:

Bridge 1

Gross soffit area 9000 SF

Average unit cost to set and strip gross soffit $1.33 per SF

Total overhang width 8 feet, $= 21.05\%$

Interior soffit width 30 feet, $= 78.95\%$

Total width of deck 38 feet, $= 100.00\%$

Bridge 2

Gross soffit area 10,000 SF

Average unit cost to set and strip gross soffit $1.35 per SF

Total overhang width 7.84 feet, $= 20.40\%$

Interior soffit width 30.60 feet, $= 79.60\%$

Total deck width 38.44 feet, $= 100.00\%$

For two similar decks, the average unit costs shown represent very close agreement, and any contractor would be pleased with such results. The preceding data may be restated algebraically as follows:

1. $0.7895I + 0.2105H = 1.33$
2. $0.7960I + 0.2040H = 1.35$

where

I = unit cost to set and strip the gross interior soffit

H = unit cost to set and strip the gross overhang soffit

Solving these equations simultaneously, we find that $I = \$1.98$ per SF and that $H = -\$1.11$ per SF. These unit costs are mathematically correct but are meaningless from a practical standpoint.

When average unit cost data are available from a number of jobs and when the contractor has an approximate idea of the magnitude of the unit cost for one of the separate activities, it is sometimes possible to derive separate unit costs for two combined activities which will closely fit the average unit costs realized on all of the jobs.

The process requires the use of good cost data, different combinations of equations, sound judgment, and some trial and error. It is only necessary or desirable to derive separate unit costs for combined activities in those cases where the separate unit costs are quite different and where the relative amounts of each of them may vary widely on different jobs.

8.1.69 The Concrete Takeoff

The concrete takeoff for plate girder bridge decks is similar to the one illustrated previously for precast concrete girder decks.

8.1.70 Soffit Forms

The formwork is based on a preliminary design. Diagonal struts are used for the 17-foot spacing since an excessive amount of heavy timber would otherwise be required.

The deck overhang formwork for the 17-foot girder spacing will be similar to that shown in Figure 8.58 except that when wide girder spacings are used, the spans are generally longer, the girders are deeper, and the deck overhangs are greater. In addition, cantilever construction is often used where spans are long, and the girders may be haunched at the bents.

8.1.71 Minor Concrete Structures

Headwalls and drainage inlets required for pipes up to about 3 feet in diameter are frequently bid under a separate job item for minor structures. The individual struc-

Figure 8.66. *Plate girder bridge, Interstate 5 Crossing U.S. 99. Contractor: Tumblin Company.*

tures included in the item will usually contain 1.5 to 2 CY of concrete each, although when many structures are involved, a few of them may require as much as 6 or 7 CY. These larger structures are frequently bid under a separate job item for structure concrete.

In most cases, the minor structures item will, in itself, constitute only a minor portion of the overall project. Thus, a simple box culvert job may contain an item for 4 CY of minor concrete structures, while a freeway project may have several hundred cubic yards in the item.

The bid price per cubic yard for minor concrete structures normally includes the costs of all required excavation, backfill, reinforcing steel, and some of the miscellaneous iron in addition to the costs for forming, placing, finishing, and curing the concrete.

Drainage inlets may require curb angles, an occasional small steel beam, or $\frac{3}{4}$-inch-diameter steel rungs. The costs to furnish and install such items may be included in the price bid for minor structures. The material is furnished, FOB job, at a separate lump sum price by the miscellaneous iron supplier and is installed by the general contractor.

The cost to furnish metal frames, covers, and grates is usually included in the job item for miscellaneous iron. This material is installed by the general contractor, and the installation cost is included in the price bid for minor structures.

Similarly, the reinforcing steel required is quoted as a separate item, FOB job, by the reinforcing steel subcontractor, and this steel is installed by the general contractor.

Figure 8.67. Minor structures.

PIPE HEADWALLS

D	H	L₁	L₂
12"	2'-8"	5'-0"	8'-0"
24	3-8	8-6	12-6
36	4-8	12-0	17-0
48	5-8	15-0	21-0

℄ COVER ⅋ FRAME

L 3½ × 3½ × ⅜

CURB BEYOND

GUTTER

DRAINAGE INLET

The cost of minor concrete structures is estimated by applying overall unit costs to the total cubic yards of concrete in the item. A good set of plans will have the concrete and reinforcing steel tabulated for each structure; no additional breakdown is required, and valuable time is saved by omitting a detailed analysis of this minor job item.

The volumes of embedded pipes and of openings are deducted in determining the pay quantities for minor concrete structures.

Typical pipe headwalls and a typical drainage inlet are shown in Figure 8.67. Drainage inlets are designated by types. Some may have curb openings as shown; others may have grates, and some may require shallow collection boxes draining into the inlets themselves.

8.1.72 Unit Costs

Unit costs for minor concrete structures are shown in Table 8.15. These unit costs include the costs of excavation and backfill, the costs for installing all miscellaneous iron and reinforcing steel, and the costs to form and place the concrete complete, including all necessary finishing and curing.

In some cases, a small drywell may be required in the bottom slab of the drainage inlets. The cost to furnish and install such drywells is not included in the unit costs shown since drywells are seldom required.

8.1.73 The Minor Concrete Structures Estimate

The estimate shown in Figure 8.68 assumes 60 CY of headwalls and 48 CY of drainage inlets are included in the item. An allowance of 8 percent is made for waste concrete since many small pours are involved and concrete is ordered in $\frac{1}{4}$-CY increments.

The allowance for form lumber depends on the job item for structure concrete. In many cases materials from this item may be reused and no additional allowance is necessary. In our estimate we assume that the plyform will be available on the

TABLE 8.15 Unit Costs for Minor Concrete Structures

| | Labor | | Rental | | |
Activity	Unit Cost ($)	Index	Unit Cost ($)	Index	Unit Of Measure
Pipe headwalls, complete	228	$K_{6+2} = 1000$	20	$K_{BH} = 25$	CY
Drainage inlets, complete	336	$K_{6+2} = 1000$	24	$K_{BH} = 25$	CY

Note: See text for discussion of the items included in these unit costs. When drainage inlets are bid as a structure concrete item, the costs will be about as follows:

SS wall forms $2.80 per SF ($K_{6+2}$ = 1000)
SS soffit forms $3.00 per SF ($K_{6+2}$ = 1000)
Pour concrete $40 per CY ($K_L$ = 13.48)

ITEM 63 108 CY MINOR STRUCTURES				
C 16.34 $K_{6+2} = 1000$ L 13.48 F 16.10 $K_{BH} = 25$		L	DAYS	M, R
	108 × 1.08 = 117 CY 30	}		3510
	2× 0.108 × 20 = 2.2 MFBM HAVE 100			220
	PLY 0.108 × 12 = 1.3 MSF NC			—
	MISC. EXP. 0.108 × 4200		29810	455
	REINF. FOB JOB 108 CY @ 31		1000 × ⅓	3350
	CURB ∠s ¢ STEPS LS		= 90	600
			90/3 =	—
60.0	60 CY HEADWALLS 228 20	13680	30 D	1200
48.0	48 CY DI 336 24	16130		1150
108.0		29810		
	INCL. PAYROLL EXP. (1.195)	29810		10485
		35620		
	$\Sigma = \$46,105$ ($427/CY)			

Figure 8.68. *Estimate for minor structures.*

job but plan to use 2-inch material from our yard. In estimating the amounts of form materials required, we use approximate factors of 20 MFBM of 2-inch material and 12 MSF of plyform per MCY of concrete involved.

The allowance for miscellaneous expenses is based on recent cost records, when available, or may be estimated if necessary. The estimated allowances for form oil, curing compound, and chamfer strip are based on a consideration of the quantities relating to one typical headwall and to one typical drainage inlet together with the total volumes of concrete involved. The total chamfer strip required would be reduced by as much as 75 percent to allow for reuse since short lengths are used.

Items from the general costs checklist previously listed on the structure concrete estimate are not duplicated here except for those items whose quantities are specifically increased because of the minor structures work.

The allowance for payroll expenses is based on workmen's compensation insurance rates for concrete work NOC.

8.1.74 Pipe Headwalls

Headwalls for pipes over 4 feet in diameter are frequently bid under a separate job item for structure concrete. The bid unit price includes the costs to form, pour, finish, and cure the headwall complete. The attendant structure excavation, structure backfill, and reinforcing steel are paid for under separate job items for this work.

8.1.75 Unit Costs

Unit costs for large-pipe headwalls are shown in Table 8.16. The cost to finish the front wall face at its junction with the pipe or pipes depends on the circumferences of the pipes and the number of pipe ends involved. The unit of measure for this unit cost is inches of pipe diameter, and this cost is in addition to the usual costs for additional finishing.

8.1.76 Miscellaneous Concrete Costs

Unit costs for a number of miscellaneous activities encountered in the construction of engineering structures are given next. These unit costs may also be useful in estimating the cost of similar work by the comparison method.

TABLE 8.16   Unit Costs for Pipe Headwalls

| | Labor | | Rental | | |
| | Unit Cost | | | | Unit of |
Activity	$	Index	Unit Cost	Index	Measure
Fine grade	0.30	$K_L = 13.48$			SF
Make wall forms	0.33	$K_{6+2} = 1000$			SF
SS foundations	2.46	$K_{6+2} = 1000$			SF
SS walls H5′–H9′	1.35	$K_{6+2} = 1000$			SF
Pour	13.70	$K_L = 13.48$			CY
Finish at pipe ends	0.42	$K_F = 16.10$			Inch[a]

[a]Per inch of pipe diameter per end.

Note: When a crane is required to place the concrete, the labor cost will remain as shown, and the production rate for the crane will be about 8 to 10 CY per hour.

8.1.77 Post Inserts

Indexed to $K_{6+2} = 1000$, 3-inch-diameter-by-12-inch-long galvanized cans or pipe post inserts were installed at an average unit cost for labor of $6.90 each.

8.1.78 Rubber Water Stop

Indexed to $K_{6+2} = 1000$, a 6-inch rubber water stop was installed at expansion joints at a unit cost for labor of $0.77 per LF.

8.1.79 Poured-in-Place Falsework Footings

Ninety-two footings 20 inches square by 8 inches deep were excavated in dry sand and poured at a unit labor cost of $7.00 each, or $102 per CY, indexed to $K_L = 13.48$.

Six hundred twenty footings 30 inches square by 15 inches deep were excavated in dry sand and poured at a unit labor cost of $12.80 each, or $44 per CY indexed to $K_L = 13.48$.

8.1.80 Wall Blockouts

Blockouts 2 feet square were placed in a form for a 12-inch wall at a unit labor cost, indexed to $K_{6+2} = 1000$, of $67 each, or $8.33 per SF of blockout form.

Concrete pipes 18 inches in diameter were inserted in the blocked-out openings and were then formed and poured into the wall at a unit labor cost, indexed to $K_{6+2} = 1000$, of $70 each.

Blockouts 18 inches square were placed in retaining wall forms 15 inches thick at a unit labor cost, indexed to $K_{6+2} = 1000$, of $64 each, or $8.53 per SF of blockout form. The blocked-out openings were later formed and poured at a unit labor cost, indexed to $K_L = 13.48$, of $20 each.

8.1.81 Bridge Curbs

Indexed to $K_{L+F} = 29.58$, 35 CY of bridge curb 2 feet wide and 9 inches high were poured, given a light trowel finish, and water cured at a unit labor cost of $24.50 per CY. One hundred CY of bridge curb 2 feet wide and 15 inches thick were poured, finished, and cured on the same job at a unit cost for labor of $23.00 per CY.

8.1.82 Slope Paving

Indexed to $K_L = 13.48$, $1\frac{1}{2}$:1 slopes with a vertical height of 12 feet were fine graded for a unit labor cost of $0.28 per square foot, and $6 \times 6/10 \times 10$ mesh was placed on the slope at a unit labor cost of $0.12 per SF.

Indexed to $K_{L+F} = 29.58$, a 4-inch-thick concrete slab was poured on the slope, finished, and cured with curing compound at a unit cost for labor of $0.51 per SF. The slab was floated with wood floats and given a light broom finish.

8.1.83 Concrete Posts

Two hundred forty-eight posts 6×9 inches in cross section by 2 feet 6 inches in height were set and stripped at a unit labor cost, indexed to $K_{6+2} = 1000$, of $10.50 each, or $4.33 per SF. Indexed to $K_L = 13.48$, the unit labor cost to pour the posts was $7.10 each, or $150 per CY.

8.1.84 Pour Sign Foundations

Indexed to $K_L = 13.48$, 12-CY sign foundations were poured directly from the transit-mix truck and water cured at a unit cost for labor of $3.10 per CY.

8.1.85 Pour with Conveyor

Indexed to $K_L = 13.48$, bridge bent columns and caps were poured with a portable conveyor at a unit labor cost of $11.00 per CY.

8.1.86 Precast Flat Slab Deck Units

Indexed to $K_{L+F} = 29.58$, a casting bed consisting of 2200 SF of concrete 4 feet wide by 4 inches thick was fine graded, formed, poured, finished, and cured at a unit cost for labor of $0.63 per SF.

Indexed to $K_{6+2} = 1000$, 4100 SF of side forms 1.67 feet in height were set and stripped at a unit cost for labor of $1.68 per SF, and 216 pcs of treated cardboard void forms 10 inches in diameter by 13 feet in length were installed and tied down at a unit cost for labor of $6.90 each, or $0.53 per LF.

Indexed to $K_{L+F} = 29.58$, 218 CY of slab units 4 feet wide by 30 feet in length were poured, finished, and water cured at a unit cost for labor of $11.80 per CY.

The units were erected with a 2-inch space between units, and this space was filled with grout. Indexed to $K_{6+2} = 1000$, 1000 LF of 2-inch-wide grout soffit was set and stripped at a unit cost for labor of $2.56 per LF. The soffits of the erected deck units were approximately 25 feet above the ground surface.

Indexed to $K_L = 13.48$, 11 CY of grout was poured, finished, and cured at a unit cost for labor of $38 per CY. The grout soffit forms were held with she-bolts at 5-foot centers, and these holes were filled, on completion, at a unit cost for labor of $5.20 each indexed to $K_L = 13.48$.

8.1.87 Seal Form Panels

Indexed to $K_L = 13.48$, large quantities of plywood form panels 8 feet square were sprayed with form sealer using a garden-type spray can at a unit cost for labor of $0.03 per SF.

8.1.88 Sheet Metal Radius Forms

The ends of the interior walls of a box culvert may be semicircular in cross section, and in such cases, the ends of the exterior walls are formed in the shape of a quarter circle. Galvanized sheet metal is used for this work, and indexed to $K_{6+2} = 1000$, the sheet metal will be set and stripped at a unit cost of $2.80 per LF.

8.1.89 Stripping Forms

Unit costs for formwork are based on the total cost to set and strip the forms. The unit cost to strip wall forms will average about 25 percent of the total setting and stripping cost.

8.2 BUILDINGS (LIGHT CONSTRUCTION)

Buildings are bid on a lump sum basis, and there is no item for structure concrete as such. A section in the specifications will cover the concrete work, but there is no engineer's estimate of the quantity involved.

The procedures to be followed in taking off and estimating the costs of concrete in buildings are identical to those illustrated for engineering structures.

As a general rule, the unit costs for structure concrete in buildings are higher than those for similar operations on engineering projects; buildings are more irregular in plan and in elevation, their sections are not as massive, and access for men and for equipment is often limited.

In this section, typical components of concrete buildings are illustrated, the quantities to be taken off are discussed, and typical unit costs are presented. The unit costs realized on building work tend to show more variation, from job to job, than do the unit costs for engineering structures.

Engineering structures are designed primarily for utility and economy and are standardized to a great extent. Architects, however, strive to make each building different from their preceding designs and to create, at a reasonable cost, a structure that is both beautiful and functional. Each building, therefore, tends to reflect the individual tastes and preferences of its designer, and each, to a varying degree, presents the contractor with a unique set of problems and required procedures.

The ability to modify the unit costs developed on past work to fit the particular conditions on a proposed project requires sound judgment and a solid background of field and office experience since the estimator must be thoroughly familiar with the job conditions relating to those unit costs being adjusted.

8.2.1 General Costs

The discussion and the general costs checklist presented at the beginning of this chapter also apply to structure concrete estimates for building work.

8.2.2 Screeds

Screeds are required for all slab pours in building work, and the rental charge for the screed supports and brackets is a general costs item. The labor cost to set screeds is included in the cost to pour the slab.

Figure 8.69. *Bakersfield Memorial Stadium, Bakersfield College. Contractor: Tumblin Company.*

8.2.3 Additional Finishing

The costs for additional finishing on building work are based on a takeoff of the quantities of work required. The estimator must, therefore, be thoroughly familiar with the requirements of the specifications and with the interior finish schedules for all rooms having concrete walls or ceilings.

8.2.4 Sandblasting

Sandblasting construction joints is seldom required on building work. Laitance is water cut from the construction joints, the dowels are washed clean after each pour, and the costs of this work are included in the unit costs to place the concrete. When sandblasting is required, the areas involved are included in the concrete takeoff.

8.2.5 Hoisting and Moving Concrete

The costs to move concrete vertically and horizontally from the point of delivery to the pour location represent a significant portion of the concrete-placing costs on building work.

8.2.5.1 CONCRETE BUGGIES. Man-powered concrete buggies or carts are frequently used to deliver concrete laterally to the pour location. Typical buggies have a water level capacity of 6 CF and will deliver about 4.5 CF of concrete per load.

Unit costs for buggying concrete are shown in Table 8.17. To achieve these unit costs, the buggies should be loaded from hoppers and must operate in continuous

TABLE 8.17 Unit Costs for Building Foundations

Activity	Unit Cost $	Index			Unit of Measure
			Labor		
FG spread and wall footings	0.28	K_L	=	13.48	SF
FG grade beams	0.32	K_L	=	13.48	SF
Make forms for foundation walls and grade beams	0.32	K_{6+2}	=	1000	SF
SS wall footings H1[a]	2.35	K_{6+2}	=	1000	SF
SS foundation walls H2–H4	2.10	K_{6+2}	=	1000	SF
SS spread footings and pedestals[a]	2.25	K_{6+2}	=	1000	SF
SS grade beam tops only [a]	4.75	K_{6+2}	=	1000	SF
SS grade beams H2–H3	1.35–$1.90	K_{6+2}	=	1000	SF
Pour foundations					
Directly from truck[b]					
Wall footings and foundation walls	8.00	K_L	=	13.48	CY
Spread footings	7.00–$10.00	K_L	=	13.48	CY
Grade beams	4.75–$ 9.00	K_L	=	13.48	CY
16″ Φ CIDH piles	4.00–$ 6.00	K_L	=	13.48	CY
Additional to set and move buggy runways	5.00–$ 9.00	K_{6+2}	=	1000	CY
Additional to buggy concrete[c]					
Up to 50 feet one way	3.75	K_L	=	13.48	CY
100 feet one way	4.50	K_L	=	13.48	CY
150 feet one way	5.25	K_L	=	13.48	CY
200 feet one way	6.00	K_L	=	13.48	CY

[a]The cost to make the forms is included.
[b]If the foundations are poured with a crane, the production rates will be about as follows:

Wall footings and foundation walls	10CY per hour
Spread footings	10-15 CY per hour
Grade beams	7-15 CY per hour
Foundation pads containing 20 CY or more of concrete	25-30 CY per hour

See Section 8.2.6 for discussion of pour costs when concrete pumps are used.
[c]Unit costs assume that hand-powered buggies are operating under average conditions, the full buggy production can be utilized, and unneeded buggies are taken off the pour as the haul distance decreases.

cycles on level runways. Loaded buggies deliver concrete on one runway and return empty on another. When flat ramps in the runway pattern are unavoidable, an additional laborer must be stationed at the ramp to assist the loaded buggies up or down.

Buggy runways are made in sections 4 feet wide by 8 feet long using 1 × 6 boards on 2 × 4 runners. These sections must usually be supported on wooden horses or light wooden bents spaced on 4-foot centers. Based on $K_{6+2} = 1000$, the runways will cost about $0.30 per SF to make and will require 1.75 MFBM of material per MSF. The cost to make the horses will be about $300 per MFBM.

When buggies are required, the number of buggies together with the estimated amount of runways to be furnished are noted in the miscellaneous column of the takeoff for the section requiring them, and the cost for furnishing these items is carried in the general costs section of the concrete estimate. The costs to set the runways and to move them as required during the pour are included in the unit costs to pour the concrete.

8.2.5.2 HOISTING CONCRETE. Tower or jumping cranes are used to hoist all materials on tall multistory buildings. Ideally, the crane or cranes will cover the entire building area, all walls and floors are poured directly with concrete buckets, and no buggies are required.

The entire cost of tower cranes is a direct job overhead expense, as discussed in Chapter 3, and no additional charges are necessary when they are used to place concrete.

Concrete hoists together with buggy hoppers at the pour elevation are often used on buildings two to five stories high. These hoists are equipped with self-dumping concrete buckets and are not suitable for use as general material or personnel hoists. Concrete hoists are normally rented on a monthly basis for the duration of the elevated concrete work. The rental cost, the on-and-off charges, the costs to set up and to dismantle the hoist and to add tower sections as the work progresses, and the costs of fuel, maintenance, and the operator are all included in the general costs section of the concrete estimate.

The labor cost for the hoist operator may be based on an average pour rate of 6.5 CY per hour for wall concrete and 12 CY per hour for floor pours, which will account for the full day's pay that is required on part-day pours. The operator will only be hired on those days that the hoist will be used. The remaining costs noted in the previous paragraph vary with the particular hoist being used, and the amount allowed is based on conversations with the supplier or manufacturer and on the estimator's judgment.

Some concrete hoists are portable and may be towed to and from the job; on work over one story high, however, they are not easily moved at the site, so the hoist is set up at a central location and the concrete is buggied to the pour.

On buildings up to two stories high, truck cranes are often used to hoist concrete, and little or no buggying is necessary.

8.2.6 Portable Concrete Pumps

In many areas, truck-mounted concrete pumps are now available (Figure 8.70). These pumps are equipped with hydraulic booms and can pump concrete to heights of 100 feet or more. The concrete is delivered to the pumps by transit-mix trucks. Rubber hose attached to the boom enables concrete to be placed directly into wall or column forms, when accessible, and no collection hoppers or elephant trunks are required. Additional sections of hose are added when the concrete must be moved laterally to areas that are inaccessible to the pump's boom. The rate of pouring will be governed by the job conditions since typical concrete pumps have ample capacity.

The equipment cost for pumping concrete is based on an hourly rental rate for the equipment, and the rental rate for a concrete pump equipped with a hydraulic

Figure 8.70. *Truck-mounted concrete pump; A. & H. Concrete Pumping, Bakersfield, California.*

boom is currently about equal to the rental rate for a 35-ton truck crane. Since the rate of pour is governed by the job conditions, the equipment cost to pour with a pump is about the same as when a truck crane is used.

When the pour location is accessible to the pump's boom, the labor cost to pour concrete walls, columns, or slabs will be essentially the same as when the concrete is placed with a crane or is poured directly from transit-mix trucks.

When the pour location is not accessible to the pump's boom, additional labor will be required to handle the hose, and the amount allowed will depend on the anticipated job conditions and on the estimator's judgment. Assuming job conditions are such that two laborers will be able to handle the hose, the labor cost to pour columns in such locations will be about 25 percent greater than to pour them directly with a crane, and the labor cost to pour walls will be about 30 percent greater. The labor cost to pour slabs or floors, under such conditions, will be about

20 percent greater than to pour them directly from transit-mix trucks or with a crane.

When concrete must be moved laterally to the pour location, the concrete pump will eliminate the need to buggy concrete and handle runways. In every case, the resulting savings will exceed the added costs of handling the hose.

8.2.7 The Concrete Takeoff

The concrete takeoffs for building work are identical to those illustrated earlier in this chapter for engineering structures. In all cases, the concrete takeoff follows the construction sequence.

Thus, section 1 of the takeoff will cover the foundations, section 2 will cover the basement walls, section 3 will cover the basement columns, and section 4 will cover the first-floor slab. This process is repeated, floor by floor, until all the concrete in the walls, columns, floors, and roof has been taken off.

The concrete stairs are then processed in a separate section in which each stair is taken off from the bottom to the top. The interior and exterior grade slabs are taken off next, and the miscellaneous exterior concrete work is processed last.

On completion, the totals for those takeoff sections that relate to similar types of work are collected and summarized in a combined takeoff. The process is identical to that illustrated earlier, where a combined takeoff for several bridges was presented.

Structure excavation and backfill are taken off as a separate operation, after the concrete and in many cases the carpentry work have been processed. Critical dimensions and elevations, determined in connection with the concrete takeoff, are penciled in on the foundation plan to facilitate the structure excavation and backfill takeoff.

8.2.8 Foundations

A portion of a typical foundation plan for a concrete building is shown in Figure 8.71 together with a section through a typical grade beam. Grade beams are often used in the foundations for buildings, although none are required in the foundation plan illustrated.

In practice, the footings, pedestals, pilasters, and all columns, floor by floor, are usually detailed in a single table on the structural plans. Reinforcing steel and the dimensions of the various sections are indicated by letter designations on typical sections, and the actual bar sizes and spacings together with the dimensions of the structural sections themselves are presented in the tabulation.

8.2.9 The Foundation Concrete Takeoff

The concrete takeoff is similar in all respects to those illustrated earlier.

Footing 2B is taken off as follows:

Fine grade	8.00 × 8.00	=	64 SF
Footing forms H1.8	8.00 × 1.83 × 4	=	59 SF
Footing concrete	8.00 × 8.00 × 1.83/27	=	4.34 CY
Pedestal forms H1.3	1.50 × 4 × 1.33	=	8 SF
Pedestal concrete	1.50 × 1.50 × 1.33/27	=	0.11 CY

In taking off footing 2A, no deduction in form area is made at the intersection of the spread footing and the wall footings, and the takeoff will be as follows:

Fine grade	5.17 × 5.17	=	27 SF
Forms H1.3	5.17 × 1.33 × 4	=	27 SF
Concrete	5.17 × 5.17 × 1.33/27 +		
	0.33 × 1.67 × 0.67 × 2/27	=	1.34 CY

The concrete calculation includes an allowance for that additional concrete required where the wall footings will be stepped down 4 inches at their intersection with the spread footing. By including this small amount of concrete here, it will not be necessary to refer back to the spread footing depth when the wall footings are taken off.

The takeoff for that length of wall footing shown in Figure 8.71 is as follows:

Fine grade	1.67 × 34.33	=	58 SF
Forms H1	1.00 × 2 × 34.33	=	69 SF
Concrete	1.67 × 1.00 × 34.33/27	=	2.12 CY

In taking off concrete walls, a vertical section together with the related plan length of wall are used. Where corners are involved, the wall lengths are taken to the centerlines of the intersecting walls. This procedure allows the single wall length determined to be used to calculate both the contact form area of the wall and the CY of concrete in it. The additional areas and concrete due to pilasters are based on separate calculations in each case.

The takeoff for the foundation wall shown in Figure 8.71 is as follows:

Wall forms H2

$$2.00 \times 2 \times 51.50 + 1.00 \times 2 \times 2 = 210 \text{ SF}$$

Concrete:

$$0.833 \times 2.00 \times 51.50/27 + 0.50 \times 1.33 \times 2.00/27 \times 2 + 0.50 \times 0.50 \times 2.00/27 = 3.29 \text{ CY}$$

In taking off the required form lumber, an allowance should be made for the additional material required to form the slab key at the top of the foundation wall since this will be a separate piece nailed inside the wall form. The cost to set and strip all solid keyways is included in the unit costs to set and strip the walls.

Figure 8.71. Building foundations.

279

Figure 8.72. Science Building, California State University, Fresno. Contractor: Tumblin Company.

In the preceding takeoffs, no allowance is made for pour bulkheads; their number is not known at this time, they constitute only a small portion of the work to be done, and their costs are included in the unit costs for the related formwork.

No allowance is made for forming the starter walls, and they are not considered when determining concrete volumes. The costs to set and strip starter walls are included in the unit costs to set and strip the related formwork, and the volumes of concrete involved are small.

Grade beams may be poured in neat excavations as shown, or they may be completely formed. The grade beam illustrated will require 1.33 SF of grade beam top to be set and stripped per linear foot of grade beam. If the grade beam were to be completely formed, 5.00 SF of grade beam forms would be required per linear foot, and the unit cost used for setting and stripping would include the cost of forming the top portion of the grade beam.

When grade beams are poured neat, the specifications may require increasing the section 2 to 4 inches in width. The additional concrete required by such specifications and by over-width excavation is included directly in the concrete takeoff. Such concrete is carried separately on engineering work to facilitate the checking of the engineer's estimated volume of concrete.

8.2.10 Unit Costs

Unit costs for building foundations are shown in Table 8.17.

The unit cost to set and strip grade beams will be low when long, unobstructed lengths are involved. When the grade beams intersect with spread footings or when short lengths with many corners are required, higher unit costs apply.

The unit costs for placing concrete include the costs of preparing for the pour, cleaning up afterward, curing during regular working hours, and placing the concrete itself.

Figure 8.73. *Kern County Jail, Bakersfield, California. Contractor: Tumblin Company.*

The low unit cost for pouring grade beams applies to wide-grade beams having little or no top restrictions. The high unit costs apply to narrow-grade beams with severe top restrictions.

A meaningful analysis of the cost to set and move buggy runways for foundation pours cannot be made at the time the estimate is being prepared because too many assumptions would be required. The unit costs shown do represent an order of magnitude for this work, but sound judgment based on experience and past job cost records must be used in assigning the probable cost for this item.

8.2.11 Walls

A partial floor plan of a beam and slab floor is illustrated in Figure 8.74 together with a typical basement wall section.

8.2.12 The Concrete Takeoff

Building walls are taken off one floor at a time, and the height of the walls depends on the floor system used for the floor above. As a general rule, the construction joint at the top of the wall is located at the soffit elevation of the deepest beam framing into it, and the bottom construction joint is at the floor line.

A vertical wall section is used to calculate the volume of the wall concrete and the contact area of the wall forms (Figure 8.75). The length used for both calculations is taken to the intersection of the wall centerlines at corners. The additional volumes and areas relating to pilasters are calculated separately.

Figure 8.74. Beam and slab floor.

Figure 8.75. Wall forms, Scaffold jack staging.

The takeoff for the wall shown from column line B to column line 2 in Figure 8.74 is as follows:

Basement wall forms H10

$$9.83 \times 2 \times 42.67 + 0.67 \times 2 \times 9.83 = 852 \text{ SF}$$

Concrete:

$$0.667 \times 9.83 \times 42.67/27 + 0.67 \times 1.33 \times 9.83/27 + 0.67 \times 0.67 \times 9.83/27 = 10.8 \text{ CY}$$

As the concrete in each length of wall is taken off, the volumes of wall openings are deducted, and the dimensions of the openings are listed in a single column on the takeoff sheet. The information in this column is used to determine the cost of the 2-inch material and of the labor required to form or block out the openings.

In general, no deduction in wall form areas is made for small openings or for doorways or windows up to 8 feet in width. In such cases the gross form area must be set and stripped, and the wood bucks used to form the blockouts entail additional work. Where window openings occur, the plywood is removed from at least one face of the forms at the opening to enable concrete to be placed and vibrated below the sill.

Where wide entryways are encountered, the spandrel wall or beam above is supported on shoring, and only the actual contact area of the spandrel wall or beam is listed. In such cases the linear feet of spandrel wall or beam soffit are listed on the takeoff and constitute an additional cost item on the estimate sheet.

Where wide window openings occur, the wall forms are set full height, although there will be no plyform on the studs at the openings. No deduction in form area is made, and the spandrel wall above is supported by cross-pieces nailed to the wall form studs. The linear feet of such spandrel wall soffit is noted in the column for opening dimensions discussed in the preceding.

Where spandrel walls occur in connection with structural slabs, it may be possible to set a keyway for the slab and to form the spandrel wall full height. In such cases the structural slab is poured after the spandrel wall is stripped; as a general rule, savings are realized when job conditions permit this method of construction.

8.2.13 Unit Costs

Unit costs for concrete building walls are shown in Table 8.18.

The cost to set and strip staging, when required, is in addition to the costs shown for setting and stripping walls, and an estimated 1.1 MFBM of staging lumber will be required per MSF of wall face covered by the staging.

TABLE 8.18 Unit Costs for Building Walls

Activity	Labor		Rental		Unit of Measure
	Unit Cost ($)	Index	Unit Cost ($)	Index	
Make form panels					
Walls (uniform)	0.25	$K_{6+2} = 1000$	—		SF
Walls (irregular)	0.33	$K_{6+2} = 1000$	—		SF
SS basement walls H10–H12	1.65	$K_{6+2} = 1000$	—		SF
SS first through fourth floor walls			—		
H8–H14	1.80	$K_{6+2} = 1000$			SF
SS first floor walls H14–H20	2.00	$K_{6+2} = 1000$	—		SF
SS one face for gunite H9	1.08	$K_{6+2} = 1000$	—		SF
SS wood or light steel staging to H40	270.00	$K_{6+2} = 1000$	—		MSF[a]
SS planter walls H3	1.71	$K_{6+2} = 1000$	—		SF
SS blockouts or bucks for openings	1500.00	$K_{6+2} = 1000$	—		MFBM
SS spandrel beam soffit W16"	6.00	$K_{6+2} = 1000$	—		LF
Pour building walls[b]	8.00	$K_L = 13.48$	10.50	$K_{Cr} = 75$	CY
Additional–set amd move runways	0.35	$K_{6+2} = 1000$	—		SF[c]
Pour planter walls[d]	14.00	$K_L + _F = 29.58$	—		CY

[a]Per MSF of wall face covered by the staging.
[b]Unit costs assume pour location is accessible to the crane and that the concrete can be deposited directly into the forms. When pours are large or average in size, the unit cost for the crane includes 1 hour of travel time. On small pours, travel time will be additional. See section 8.2.6 for discussion of pour costs when concrete pump is used.
[c]Per SF of runway area.
[d]Costs to hoist or buggy concrete, when required, are additional.

The unit costs shown for setting and stripping wall forms include the costs of handling scaffold jacks. The rental charges for the scaffold jacks are carried in the general costs section of the concrete estimate.

The unit cost shown for setting and stripping wall blockouts is based on the total amount of lumber required including the interior braces and struts. In practice, the estimator determines the cost to block out typical door, window, and duct openings, and the list of openings from the takeoff is reduced to an estimated number of typical openings for pricing purposes.

Building walls must usually be poured before setting the soffit forms for the floor next above. When concrete buggies are used, the cost to set and move the elevated runways represents a substantial portion of the pouring cost and demonstrates the advantages of using tower cranes or concrete pumps when tall buildings with large volumes of wall concrete are involved.

The cost to set and move buggy runways may be based on an estimated area of runways to be furnished, and this varies with the layout of the walls.

8.2.14 Form Lumber

On small buildings, it may be necessary to make wall form panels for all the walls since little or no reuse may be possible. Forms must be made for a minimum of 2 $\frac{1}{2}$ wall pours, and in most cases at least 5 MSF will be required.

On large buildings, the area of wall forms required may be approximated by dividing the total contact area of all the walls by an estimated 6 reuses for the forms. Thus, materials will be furnished for and the unit cost for making forms

Figure 8.76. *Probation Department Building, Bakersfield, California. Contractor: Tumblin Company.*

applied to one-sixth of the total wall area. It will be required to use 1.17 MSF of plyform and 2.0 MFBM of lumber MSF of wall form panels furnished.

An additional unit cost to cover the labor required to modify the forms for reuse must be applied to the remaining five-sixths of the total contact area of the walls. This unit cost may amount to 30 percent of the unit cost used to make the forms, and the percentage used will depend on the amount of variation in the successive wall pours and on the estimator's judgment.

8.2.15 Columns

Unit costs for columns are shown in Table 8.19.

Columns are poured from the floor soffit forms before installation of the floor-reinforcing steel (Figure 8.77). When concrete buggies are used, no allowance for setting or moving runways is necessary.

Column pours entail small amounts of concrete, which is placed in three or more lifts to prevent excessive form pressures. Hoisting costs are reduced when columns are poured in conjunction with other larger pours, in which case the hoisting cost for the larger pour governs.

Excessive fluid pressure in sonotube forms will result in higher costs for additional finishing and for stripping. When high lifts are combined with maximum vibration, the unit cost to set and strip sonotubes may run as high as $8.00 per LF based on $K_{6+2} = 1000$. When sonotubes are used, the pouring and stripping operations should always be properly supervised.

Since columns are formed in conjunction with the floor soffit forms, the amount of column forms required will depend on the area of floor forms furnished. No allowance for modifying column forms for reuse is necessary.

8.2.16 Beam and Slab Floors

A partial plan of a beam and slab floor is shown in Figure 8.74, and typical formwork is shown in Figure 8.79.

TABLE 8.19 Unit Costs for Building Columns

Activity	Labor Unit Cost ($)	Index	Rental Unit Cost ($)	Index	Unit of Measure
Make form panels	0.35	$K_{6+2} = 1000$	—		SF
SS columns (16x16) H8–H12	2.10	$K_{6+2} = 1000$	—		SF
SS sonotube (16″ Φ) H8–H12	4.00	$K_{6+2} = 1000$	—		SF
Pour with crane	12.00	$K_L = 13.40$	28		CY

Note: The unit costs for pouring assume normal access for the crane, and the concrete can be deposited directly into the forms. When 11 CY or more of columns are poured at one time, the unit cost for the crane will include 1 hour of travel time. See Section 8.2.6 for discussion of pour costs when concrete pump is used.

Figure 8.77. *Column pour. (Photograph by The Burke Company.)*

8.2.17 The Concrete Takeoff

Structural plans designate concrete beams by width and depth of the beam cross section. In all cases the width is listed first; the depth is always taken from the beam soffit to the top of the floor slab.

Beams 109 are 18 × 20-inch beams having 2.67 SF of beam sides per LF; beam 143 is a 24 x 20-inch beam having 2.67 SF of beam sides per LF.

The gross soffit area to which the unit costs are applied includes the areas of both the slab and the beam soffits. Thus, the gross soffit area enclosed by the building walls and by column lines B and 2 is 18 × 24 = 432 SF; no deductions are made for the plan areas of the pilasters or columns.

We designate that exterior wall formwork located between the upper two construction joints shown on section 1, Figure 8.74, as wall extension. This formwork

Figure 8.78. *Underground garage, beam and slab construction. Contractor: Tumblin Company.*

must be set and stripped in conjunction with the formwork for the floor. The area of the wall extension forms shown on section 1 amounts to 1.67 SF per LF of wall. The formwork required for the interior face of the wall between the same two construction joints is included with the areas of the beam sides and amounts to 1.33 SF per LF of wall.

The volumes of openings in the floor slab are deducted from the gross volume of concrete at the time the areas of the edge forms required for the openings are taken off.

The takeoff sheet is set up in a similar manner to those shown in Section 8.1. There will be separate columns for at least the following items:

Figure 8.79. *Forms for beam and slab floor.*

1. Miscellaneous.
2. Square feet of deck or deck opening edge forms.
3. Linear feet of beams.
4. Number of form ties when required.
5. Square feet of wall extension.
6. Square feet of beam sides.
7. Gross soffit area.
8. Cubic yards of concrete in the beams and slab combined.

Items such as curbs, locker bases, and cement base are included in the floor concrete takeoff when they occur. Unit costs for these items are discussed in Section 8.2.32.

8.2.18 Unit Costs

Unit costs for beam and slab floors are shown in Table 8.20.

The unit costs shown for setting and stripping beam sides and wall extensions include the costs of making the forms. The unit cost shown for setting and stripping

TABLE 8.20 Unit Costs for Concrete Floors

| Activity | Labor | | Unit of Measure | Production Rate (CY per hour) |
	Unit Cost ($)	Index		
SS wall extension	3.10	$K_{6+2} = 1000$	SF	—
SS deck edge and blockouts	2.30	$K_{6+2} = 1000$	SF	—
SS slab pour bulkheads	2.60	$K_{6+2} = 1000$	SF	—
Beam and slab floors				
SS beam sides	2.10	$K_{6+2} = 1000$	SF	—
SS gross soffit forms $H_c = 10^a$	1.65	$K_{6+2} = 1000$	SF	—
Pour and trowel finish [b]	16.00	$K_L + _F = 29.58$	CY	12
Concrete joists with steelforms				
SS falsework[c]				
Pour and trowel finish	15.00	$K_L + _F = 29.58$	CY	15
Structural slabs				
SS soffit forms $H_c = 10^d$	1.65	$K_{6+2} = 1000$	SF	
Pour and trowel finish[b]	0.40	$K_L + _F = 29.58$	SF	10
	5.00	$K_L = 13.48$	CY^e	

[a]Gross soffit area = area of beam soffits plus area of slab soffits = total area of beam and slab floor. Increase unit cost 4.5 percent per foot of additional height to $H_c = 14$.
[b]The unit costs for pouring concrete assume the concrete is poured directly from transit-mix trucks or is deposited directly into the forms with a crane or a concrete pump. The costs to hoist the concrete or to buggy or pump it horizontally are additional. Included are the costs to set screeds, pour complete, cure, clean up, and cover the slab with building paper on completion.
[c]Use the bridge falsework curve, Figure 8.17, and increase these costs by 7 percent.
[d]Increase unit costs 4.5 percent per foot of increased height to $H_c = 14$.
[e]Per CY in excess of 12.3 CY per MSF of slab.

the gross soffit includes the cost to set and strip all required falsework and soffit plyform.

The unit cost shown for pouring includes the cost of a steel trowel finish and assumes that a rotary finishing machine is used. When only a wood float finish is required, the pouring cost will be reduced by about $1.50 per CY based on $K_F = 16.10$.

8.2.19 Form Lumber

The amount of falsework and slab joists required is based on a preliminary design for a typical section of the floor. Per MSF of beam sides, 1.5 MFBM of 2 x 4 is allowed. About 1.17 MSF of plyform will be required per MSF of theoretical plyform area, and this area will be greater than the gross soffit area when the plyform for the beam soffits extends past the sides of the beams.

In general, about 2.6 MFBM of falsework lumber will be required per MSF of gross soffit area for height to concrete soffit $H_c = 10$, and the amount will increase about 4 percent per foot of increased height to $H_c = 14$. Similarly, the amount of plyform required will be about 1.3 to 1.5 MSF per MSF of gross soffit area.

In most cases, form materials are furnished for the largest floor, complete, and this material is reused on the remaining floors. Depending on the areas and number of the succeeding floors, on the soffit finish required, and on the variations in beam sizes and spacings, the estimator may allow additional plyform for some of the additional floors.

Reshoring is usually required for one or more floors immediately below the floor being formed. The amount of reshoring will depend on the load from the floor being poured and on the design live loads for the floors below. The live load assigned to the floor being poured will depend on the estimator's judgment, and 20 PSF has been used.

Typical reshoring may consist of 4 x 4 posts at midspan for each beam and under the girders at the points where the beams frame into them. Reshoring is set after the floor has been stripped, and the material required is normally in addition to the falsework required for the floor forms. Wedges must be furnished for each post, and the number of wedges is based on the MFBM of posts required.

8.2.20 Concrete Joist Floors

Concrete joist floor systems constructed with steel forms on wood falsework are frequently encountered. One-way concrete joist floors are similar to beam and slab construction in appearance. Two-way concrete joist floors are built with steel "dome" forms, and the concrete joists run in both directions on 2- to 5-foot centers. Both types of floors are, in effect, deep slabs with voids in the bottom face that are created by the steel forms. Such floors are often referred to as pan slabs or dome slabs.

Typical steel forms for concrete joist construction are shown in Figure 8.80, and a typical section through a one-way concrete joist floor is shown in Figure 8.81

Figure 8.80. Details for Ceco Adjustable Steelforms (The Ceco Corporation).

together with the required wood forms and falsework. The volumes of the voids created by various sizes of typical steel forms are given in Table 8.21.

The steel forms shown in Figures 8.80 and 8.81 are adjustable. The pans pass by and are nailed to the sides of the joist soffit forms to attain the desired joist depth. Fixed depth pans are sometimes used. These have outstanding flanges at the bottom edge that rest on and are nailed to the joist soffit material.

The steel forms or pans are normally furnished, set, and stripped by subcontractors at a unit price per square foot of gross slab area. These subcontractors usually

STEELFORM ENDCAPS

24"

3½"

18"

FORM TIE

5/8 PLY

TOT 5 - 2×4

4 × 4

TOT 2 - 1×6

4×4-4'C-C

3½"

4"

STEELFORMS

4×6-4' C-C

4×4 - 4' C-C

3'-0" TYP.

6"

6×4

3'-0"

4" MAX.

2×4 16"C-C

4×3

2×4 4' C-C

2"

CJ

CJ

5/8 PLY

2×4 12" C-C

2×4 WALES

FORM TIE

ONE-WAY CONCRETE JOISTS

ADJUSTABLE STEELFORMS

TYPICAL FORMWORK DETAILS

Figure 8.81. Concrete joist floor.

TABLE 8.21 Voids Created by Various Size Ceco Adjustable Steelforms[a]

Depth of Steelform (inches)	Cubic Feet of Voids per Linear Foot for Various Widths of Adjustable forms				
	40"	30"	20"	15"	10"
		Standard			
6		1.207	0.790	0.582	0.371
8		1.623	1.068	0.790	0.512
10		2.040	1.345	0.998	0.651
12		2.457	1.623	1.207	0.790
14		2.873	1.901	1.415	0.929
		Deepset[b]			
12	3.227	2.394	1.561	1.144	0.727
14	3.783	2.811	1.838	1.352	0.866
16	4.338	3.227	2.116	1.561	1.005
18	4.894	3.644	2.394	1.769	1.144
20	5.450	4.061	2.672	1.997	1.283
22	6.005	4.477	2.950	2.186	1.422
24	6.561	4.894	3.227	2.394	1.561

[a]These data apply to Ceco Adjustable Steelforms for one-way joist floor construction and are furnished through the courtesy of the Ceco Corporation.
[b]The tapered endform for 40" wide deepest adjustable forms is 3 feet long and tapers to 34 inches in width. See also Figure 8.80.

quote each job two ways. One quotation includes the costs of furnishing, setting, and stripping the pans on the general contractor's falsework. The second quotation includes the cost of doing this work on falsework that is furnished, set, and stripped by the subcontractor.

8.2.21 The Concrete Takeoff

On a typical concrete joist floor system, the bulk of the pans used are of uniform width. To fit the voids to the plan dimensions of the slab, however, it is usually necessary to use one or two rows of narrower pans at each side, and sometimes some of the joist widths are varied for the same reason. The volume of concrete in the floor is determined by deducting the volumes of the voids created by the pans from the volume of the solid, full-depth slab.

Wall extensions and some beam sides must always be formed in connection with pan slab work. In Figure 5.5, 2.00 SF of wall extension are required per linear foot of wall. In addition, 0.54 SF of beam sides per linear foot of wall are required for the interior face of the wall, and 1.08 SF of beam sides are required per linear foot of the 24-inch-deep girder.

8.2.22 Unit Costs

The unit costs for concrete joist floors are shown in Table 8.20. The unit costs for setting and stripping wall extensions, slab edge forms, and pour bulkheads are the same for beam and slab construction and for concrete joist floor systems. The unit cost to set and strip beam side forms is higher in the case of concrete joist floors because small areas are usually involved. For situations like those shown in Figure 8.81, the unit cost to set and strip the beam sides is about 1.5 times the unit costs shown for beam sides in connection with beam and slab floors.

The unit cost to pour concrete joist floors is less than the unit cost to pour beam and slab floors because the slab and joists are normally poured in one pass. When a wood float finish is required, the pour cost may be reduced by $0.03 per SF of gross slab area based on $K_F = 16.10$.

8.2.23 Form Lumber

The amount of falsework required is based on a preliminary design for a typical section of floor and includes the posts and beam system necessary to support the beam forms at the interior faces of the walls.

In general, about 7.0 MFBM of falsework lumber will be required per MLF of joists and beams for one-way concrete joist construction when $H_c = 10$. This quantity should be increased 4 percent per foot of increased height to H_c-14. Two-way concrete joist systems require about 2.3 MFBM of falsework per MSF of gross slab area when $H_c = 10$, and this quantity should also be increased 4 percent per foot of increased height to H_c-14.

8.2.24 Structural Slabs

Structural slabs vary greatly in area and depth, and the amount of falsework required is based on a preliminary design for a typical section in all cases.

In general, about 2.2 MFBM of lumber will be required per MSF of soffit area for H_c-10, and this quantity increases about 4 percent per foot of increased height to H_c-14. Per MSF of soffit area, 1.17 MSF of plyform will be required.

Unit costs for structural slabs are shown in Table 8.20 and are based on slab areas of 800 SF or greater per slab. The unit costs to set and strip soffit forms for small slabs are higher and may amount to as much as $3.00 per SF for slabs 50 to 100 SF in area based on $K_{6+2} = 1000$.

8.2.25 Stairs

A section of a typical concrete stair and the required formwork are shown in Figure 8.82. Stairwell walls are treated as typical building walls and are not included in the stair estimate since the stairs are constructed after these walls and the floors on them have been completed.

Material from wall and floor formwork will be available for reuse in many cases, and no additional allowance for stair soffit form materials may be necessary. In general, about 3.2 MFBM of soffit form lumber will be required per MSF of stair, or stair plus landings, plan area.

In all cases, 2-inch material must be furnished for the riser forms, and these

Figure 8.82. Stair forms.

forms should be firmly secured since they must support the workmen during the pour. In addition, the riser forms are subjected to horizontal forces resulting from the fluid pressure of the concrete.

The support details for the riser stringers will vary to suit the existing conditions. The stringers are supported with removable screed-type supports set square with the soffit on 6- to 8-foot centers. The stringers must be restrained horizontally and vertically to prevent upward rotation. Temporary struts to previously poured concrete are often used for these purposes.

When stairs are poured in a single operation, as opposed to being poured rough and topped later, the riser forms are removed during finishing, and the cost of this work is included in the pour costs.

8.2.26 Unit Costs

Unit costs for stairs are shown in Table 8.22.

The unit cost for setting and stripping stairs with no landings assumes single flights formed between previously poured slabs or beams. The unit cost for placing stair topping includes the cost of setting and stripping temporary riser forms and is applied to the total area topped, including the area of the risers.

8.2.27 Tilt-up Walls

In this type of construction, the walls are cast in a horizontal position at the site and are tilted up and moved onto previously poured foundations with a truck crane (Figure 8.83). After the panels have been erected and braced, columns or pilasters are cast in place between them. In the following discussion of tilt-up panels, the terms ''vertical,'' ''horizontal,'' and ''above'' will refer to positions on the erected panels.

TABLE 8.22   Unit Costs for Concrete Stairs

	Labor		Unit of Measure
Activity	Unit Cost ($)	Index	
SS soffit forms			
Stairs with no landings	3.50	$K_{6+2} = 1000$	SF[a]
Stairs plus landings	3.95	$K_{6+2} = 1000$	SF[a]
SS riser forms	2.35	$K_{6+2} = 1000$	LF
SS Hl stair edge	3.45	$K_{6+2} = 1000$	SF
Pour and finish complete[b]	90.00	$K_L + _F = 29.58$	CY
Pour rough[b]	30.00	$K_L + _F = 29.58$	CY
Place and finish topping[b]	3.75	$K_L + _F = 29.58$	SF
Place and finish concrete in metal pans[b]	1.20	$K_L + _F = 29.58$	LF
Set 3″ metal tread nosing	1.60	$K_L + _F = 29.58$	LF
Additional for tread abrasives	0.15	$K_L + _F = 29.58$	SF
Set 3″ sleeves for rail posts	1.85	$K_{6+2} = 1000$	Each
Set pipe stair railing			
Grouted posts	27.00	$K_{6+2} = 1000$	Each
Brackets (2 bolts ea.)	21.00	$K_{6+2} = 1000$	Each

[a]Stair soffit area is based on slope lengths.
[b]Costs to hoist and to buggy concrete are additional. Estimated production rate is 2 CY per hour.

8.2.28   General Costs

The following general cost items must be considered in addition to those earlier presented. The rental and material costs for these items are carried in the general costs section of the concrete estimate, but the installation costs, if any, are carried in the operating sections of the estimate.

8.2.28.1   CASTING SITE. It may be necessary to acquire additional space for a casting site. Maximum economy is achieved when the panels are cast on exterior grade slabs which will become part of the finished construction. In such cases, the panels are cast with their bottom edges about 5 feet from and parallel to their erected location. The truck crane is located inside the building, and each panel can be tilted up and erected without moving the crane during the process.

Since the wall foundations must remain exposed until the panels are erected, a construction joint in the exterior slab is required; it will be parallel to and about 3 feet from the wall line.

When exterior slabs are not available, wall panels are cast on the interior grade slab if space permits, and the crane is located outside the building. As before, a construction joint will be required in the floor slab near the wall line.

Space is usually limited, and panels are frequently cast in stacks of two or three panels. The stacks should be low enough to permit the concrete to be placed directly from the transit-mix truck. When panels are stacked, the locations of dowels and

Figure 8.83. *Tilt-up building. Panels 5 ½ inches thick by 38 feet in height. Contractor: Tumblin Company.*

wall openings as well as the plan dimensions of the panels themselves must be compatible with this procedure.

As a last resort, wall panels may be cast on temporary, 4-inch-thick waste slabs. The costs to furnish and remove the waste slabs are included in the operating section of the estimate.

Casting bed slabs must be smooth, uniform, plane surfaces. Small imperfections permissible in grade slabs may be objectionable when they appear in erected walls. Wall panels should not be cast over cracks, construction joints, or temporarily filled expansion joints since this will usually result in cracked wall panels.

8.2.28.2 ENGINEERING. The wall panel thickness and the reinforcing steel shown on the structural plans will have been designed to enable the panel to function as a portion of a building wall. The procedures required to cope with the additional stresses encountered during tilting are the responsibility of the general contractor.

The necessary engineering services are provided by the suppliers of the concrete accessories used in tilt-up construction. These firms submit quotations covering the costs of their engineering services and the material and rental costs for the accessories required.

During the erection process, the wall panels are tilted upward to a nearly vertical position about an axis of rotation located at their bottom edges. They are then lifted and placed in their final position on the building foundations.

As a panel is tilted, it is supported by the casting slab at its bottom edge and by crane rigging attached to from two to sixteen lift points located on the face of the panel. The lifting shackles, or lift hardware, are bolted or pinned to inserts which have been cast in place in carefully designed locations.

The lift point inserts are located to keep the bending stresses, which occur during erection, within allowable limits. The inserts are located in horizontal rows of two or four inserts each, and from one to four rows will be required. When but one row is required, two, three, or four inserts may be used, and these are sometimes located in the top edges of short panels.

A panel requiring four rows of two inserts each is designated as a 2-wide by 4-high pickup or as a 4-row, 8-point pickup. A 2-wide by 4-high pickup is illustrated in Figure 8.84. The tabulated lift point dimensions will be varied slightly by different designers.

For design purposes, the vertical elements of the panel are assumed to be continuous beams supported by the ground reaction at the bottom and by assumed transverse beams which are coincident with the horizontal rows of inserts. These horizontal rows are spaced vertically to equalize the bending moments as much as possible.

The transverse beams are supported at the lift points, and the inserts in the horizontal rows are spaced to equalize the bending moments in the transverse beams. A mat of No. 4 bars is frequently required in the upper surface of the slab at each horizontal row to increase the effective width of the transverse beam and the resistance of the inserts in tension and in shear.

TYPICAL LIFT-POINT DIMENSIONS		
	2 - WIDE	4 - WIDE
A	0.20 W	0.11 W
B	1 @ 0.60 W	3 @ 0.26 W
	1 - HIGH	2 - HIGH
C	0.29 L	0.18 L
D	1 @ 0.71 L	2 @ 0.41 L
	3 - HIGH	4 - HIGH
C	0.13 L	0.10 L
D	3 @ 0.29 L	4 @ 0.225 L

Figure 8.84. *Rigging for tilt-up panels.*

The center of gravity of the lift insert load pattern must be located on a vertical line through the center of gravity of the wall panel and should be at least 18 to 24 inches above the center of gravity of the panel. In determining the latter, the areas of all wall openings must be considered. A single rigging system should be used for all panels when possible. The cost of any additional inserts that may be required will be small compared to the cost of rerigging the crane during erection.

Typical wall panels are $5\frac{1}{2}$ to $6\frac{1}{2}$ inches thick and are too thin to function as reinforced concrete during the erection process. Paradoxically, the panels are considered as unreinforced plain concrete, the erection stresses are calculated by the flexure formula, and the working stress used is a percentage of the modulus of rupture for the concrete.

In all cases, test cylinders should be taken from the wall panel concrete, and when facilities are available, test beams should be cast to determine the modulus of rupture directly.

Crushed aggregates normally produce stronger concrete, and when used, the ratio of the modulus of rupture to the compressive strength is generally greater.

The following rules of thumb should be considered in tilt-up panel designs.

1. Lift point inserts should never be spaced more than 13 feet apart or less than 18 inches from the edges of openings or of the panels.
2. The length of the cantilever overhangs at the top and sides of panels should not exceed $t - 1$ feet, where t = the thickness of the panel in inches.
3. The number of lift point inserts used on a given panel should not be less than the total panel weight divided by the lesser of the safe working loads for the inserts in tension or in shear.

8.2.28.3 HIGH-STRENGTH CONCRETE. Concrete of a higher strength than that specified may be required to attain the necessary modulus of rupture or to attain it at an earlier age.

8.2.28.4 CURE AND BOND BREAKER. The same compound must be used to cure the casting bed concrete and to break the bond between it and the wall panel. This material is furnished by the concrete accessories supplier and must be carefully applied as directed. Two coats are required, and the material is applied with sprays at a rate of about 500 SF per gallon per coat. The second coat is applied the day before the reinforcing steel is placed, and the surface must be checked and touched up, if necessary, after the steel has been installed. Polyethylene sheets or paper are generally unsatisfactory bond breakers; wrinkling will occur in most cases.

8.2.28.5 ADDITIONAL REINFORCING STEEL. In addition to the mats that may be required at the transverse beams, extra reinforcing steel may be necessary at the sides, top, or bottom of large or unsymmetrically placed wall openings to resist erection stresses. This additional reinforcing is not included in the quotations received from reinforcing steel subcontractors, and the costs to furnish and install it are included in the general costs section.

8.2.28.6 STRONGBACKS. Strongbacks are often required when large openings occur in the wall panels and may be located in horizontal or vertical positions. They ordinarily consist of pairs of timber beams or steel channels tightly bolted to inserts cast into the wall panels. Strongbacks are designed to take all the bending moments in a given area, and the bending resistance of the wall panel concrete in that area is neglected. After the panels have been erected, the strongbacks may be removed for reuse.

8.2.28.7 SHIMS. To ensure that the bottom edges of the erected panels are set level, at least two sets of steel or masonite shims are required under each panel.

8.2.28.8 TEMPORARY BRACES. At least two braces are required for each of the wall panels to plumb and support them after erection. The bracing system is designed to withstand the anticipated wind loads and remains in place until the floor or roof diaphragms have been installed. Long braces require strut and tie supports at their midpoints.

Telescoping tubular steel braces having screw jacks for easy adjustment are furnished by the concrete accessories suppliers on a monthly rental basis. These braces are available in lengths ranging from about 16 to 40 feet. Steel end hardware with screw jacks attached is also available for use with double 2 x 4 or 2 x 6's furnished and fabricated by the contractor. When panels as high as 40 feet must be erected, a bracing system of wire rope, turnbuckles, and deadmen may be used in addition to the rigid braces.

For maximum economy, it is essential that the panel casting and erection sequence selected permit the upper ends of all braces to be bolted to the horizontal panels prior to their erection and that the grade slab inserts be clear and unobstructed when needed.

The labor cost to connect the braces to the horizontal panels and the cost to remove the braces on completion are based on a single unit cost for both operations. The labor cost to bolt the braces to the grade slabs and to install struts when required are included in the costs to erect the panels.

8.2.28.9 INSERTS AND BOLTS. This item covers the costs to furnish and install the inserts required for the lift hardware, the strongbacks, and the temporary braces.

8.2.28.10 LIFT HARDWARE. This item covers the rental charges for the lift plates and eyes that will connect to the lift point inserts. Twist-lock lift hardware is available and, when used, is required for only one panel. When bolted lift hardware is used, it is necessary to furnish hardware for two or three panels to prevent delays during erection. The costs to install and to remove lift hardware are included in the erection costs.

8.2.28.11 FILL AND PATCH INSERT HOLES. The holes at all inserts must be filled and patched on completion, and the costs for this work are carried in the additional finishing item.

8.2.28.12 DRY PACK AT FOUNDATION. This item covers the costs to furnish and place dry pack material between the bottom edges of the wall panels and the wall foundations. The costs of materials and the labor are carried in the general costs section of the concrete estimate.

8.2.29 Form Lumber

In most cases, the only new material required is the 2-inch-thick lumber used for the edge forms. The bracing necessary to align and to support these forms is normally available at the job site.

8.2.30 Unit Costs

Unit costs for tilt-up wall panels are shown in Table 8.23.

The unit cost for pouring the waste slabs and the wall panels assumes that concrete is delivered directly to the slabs in transit-mix trucks.

The unit cost and the production rate shown for pouring the columns represent an order of magnitude for this work, but good judgment must be used in selecting the unit costs to be used. Columns for tilt-up work are often small in cross section and contain a large amount of reinforcing steel; in addition to the typical column

TABLE 8.23 Unit Costs for Tilt-up Panels

Activity	Unit Cost ($)	Index	Unit of Measure
Waste slabs			
FG	0.06	$K_L = 13.48$	SF
SS edge forms H4″	0.75	$K_{6+2} = 1000$	LF
Pour and trowel finish[a]	0.32	$K_L + _F = 29.58$	SF
Wall Panels			
SS edge forms (solid)	1.15	$K_{6+2} = 1000$	LF
SS edge forms (2-piece)	1.40	$K_{6+2} = 1000$	LF
SS blockout forms	1.75	$K_{6+2} = 1000$	LF
Set inserts	5.00	$K_{6+2} = 1000$	Each
Pour and trowel or broom finish[a]	0.40	$K_L + _F = 29.58$	SF
	5.00	$K_L + _F = 29.58$	CY[b]
Columns (poured between panels)			
Make forms	0.40	$K_{6+2} = 1000$	SF
SS column forms	3.10	$K_{6+2} = 1000$	SF
Pour columns[c]	20.00	$K_L = 13.48$	CY
Set shims	1.90	$K_{6+2} = 1000$	Each
SS temporary braces	11.00	$K_{6+2} = 1000$	Each
Dry-pack panels at bottom	35.00	$K_L = 13.48$	CF

[a]These unit costs assume the concrete is poured directly from transit-mix trucks.
[b]Per CY in excess of 18.5 CY per MSF of wall panel area.
[c]The cost to hoist the concrete is additional. Estimated production rate is about 2.5 CY per hour.

bars and ties, the horizontal wall panel reinforcing projects into the columns, and these bars may have hooked ends.

8.2.31 Erecting Tilt-up Panels

Wall panels should be erected by a rigging company experienced in this type of work. Such firms are normally paid on the basis of hourly rental rates for the truck crane and for the riggers. The crane rental rate includes all required spreader beams and rigging, and the crane used should have twice the theoretical lifting capacity required by the actual job conditions. As shown in Figure 8.84, continuous wire rope rigging is used to equalize the lift point loadings, insofar as possible, during erection. See Figure 8.85.

Panels erected with the crane positioned inside the building should be cast with their inside faces up; panels cast on the building floor slab and erected from the outside should be cast with their outside faces up. This procedure obviates the deadman's pick, which occurs when a panel would fall toward the crane if the rigging were to fail during the erection process.

Two to four extension ladders are needed to enable the workmen to disconnect the rigging from the erected and braced panels.

In rare instances, the bond breaker may not perform perfectly. In such cases a slight horizontal movement induced by a good hydraulic jack may be required.

Under optimum conditions, solid rectangular 2-wide, 2-high pickups may be erected at a rate of three panels per hour, and solid 4-wide, 4-high pickups may be erected at a rate of two panels per hour. Large panels irregular in shape requiring strongbacks and additional care in handling may require 1 hour for erection.

A minimum erection crew for 2- to 4-point pickups may consist of a rigger foreman, two riggers, and two carpenters. When 8- to 16-point pickups are in-

Figure 8.85. Tilt-up panel erection. (Photograph by The Burke Company.)

Figure 8.86. Grade slabs and miscellaneous items.

volved, two additional riggers should be used. When strongbacks must be removed for reuse, two additional carpenters and, in some cases, an A-frame truck are required for only that time necessary to do this work.

8.2.32 Grade Slabs and Miscellaneous Items

Typical details are shown in Figures 8.86 and 8.87. Unit costs for this work are given in Tables 8.24 and 8.25.

The cost to pour concrete slabs is based on two separate and distinct cost factors: the cubic yard cost to place the concrete and the square foot cost to finish the surface of the slab. The unit cost per square feet shown in Table 8.24 includes the costs to place and finish the concrete in a 4-inch-thick slab, which contains 12.35 CY per MSF.

Figure 8.87. Miscellaneous building concrete.

TABLE 8.24 Unit Costs for Grade Slabs and Curbs

Activity	Labor Unit Cost ($)	Labor Index	Rental Unit Cost ($)	Rental Index	Unit of Measure
FG for specification fill	0.06	$K_L = 13.48$	—		SF
FG for concrete slabs	0.08	$K_L = 13.48$	—		SF
Place specification fill	6.50	$K_L = 13.48$	4	$K_{BF} = 31^a$	CY
SS edge forms for slabs or walks	2.25	$K_{6+2} = 1000$	—		SF
Set 6x6/10x10 wire mesh	75.00	$K_{6+2} = 1000$	—		MSF
Place vapor barrier	45.00	$K_{6+2} = 1000$	—		MSF
Hand excavation at scoops	14.00	$K_L = 13.48$	—		CY
Pour and finish grade slabs^b	0.33	$K_{L+F} = 29.58$	—		SF
plus	4.10	$K_L = 13.48$	—		CY^c
Set and move buggy runways	0.06	$K_{6+2} = 1000$	—		SF^d
Pour and finish sidewalks	0.36	$K_{L+F} = 29.58$	—		SF
plus	4.50	$K_L = 13.48$	—		CY^c
SS, P & F stud wall curbs	4.00	$K_{L+F} = 29.58$	—		LF
SS, P & F 8" curb and gutter	4.80	$K_{L+F} = 29.58$	—		LF
SS, P & F 8" curb (straight)	2.70	$K_{L+F} = 29.58$	—		LF
SS, P & F 8" curb (irregular)^f	5.00	$K_{L+F} = 29.58$	—		LF

$^a K_{BF}$ = sum of the hourly rental rates for a $\frac{1}{2}$-CY wheel loader and a hand-operated vibratory compactor.

^{b}These unit costs assume the concrete is placed directly from transit mix trucks. The cost to buggy or pump concrete horizontally is additional. The unit costs include the costs for setting screeds, pouring complete with a trowel finish, curing, cleaning up, and covering the slab with building paper.

^{c}Per CY in excess of 12.3 CY per MSF of slab.

^{d}Per SF of slab area.

^{f}Similar to footnote b except that walk is not covered with paper.

Such as curbs for parking lot islands.

Note: Curb and gutter details are shown in Figure 8.86.

TABLE 8.25 Unit Costs for Miscellaneous Items

Activity	Labor Unit Cost ($)	Index			Unit of Measure
Pipe trenches					
Fine grade	0.23	K_L	=	13.48	SF
Make, set, and strip inside face	3.30	K_{6+2}	=	1000	SF
Pour and finish, including slab	32.00	K_{L+F}	=	29.58	CY
Window sills					
Set and strip	4.00	K_{6+2}	=	1000	LF
Pour and finish	4.00	K_{L+F}	=	29.59	LF
Locker base					
Set and strip	2.45	K_{6+2}	=	1000	SF
Pour and finish	19.00	K_{L+F}	=	29.58	CY
Cement base					
6″ straight, including wall chase	2.50	K_F	=	16.10	LF
6″ cove, including wall and	6.00				
floor chases		K_F	=	16.10	LF
Manholes[a] FG, SS, and P complete	150.00	K_{C+L+F}	=	45.92	CY
Fine grade	0.28	K_L	=	13.48	SF
Set and strip bottom slab	2.30	K_{6+2}	=	1000	SF
Make wall forms	0.25	K_{6+2}	=	1000	SF
Set and strip walls H9–H12	2.00	K_{6+2}	=	1000	SF
Set and strip soffit form	3.00	K_{6+2}	=	1000	SF
Pour and finish slabs	25.00	K_{L+F}	=	29.58	CY
Pour 12″ thick walls	12.00	K_L	=	13.48	CY
Pour 8″ thick walls	22.00	K_L	=	13.48	CY
Eyebrows					
SS soffit forms H_c4^c	1.85	K_{6+2}	=	1000	SF
SS soffit forms H_c10^c	2.70	K_{6+2}	=	1000	SF
Pour and finish slabs	31.00	K_{L+F}	=	29.58	CY
Make forms for fins	0.35	K_{6+2}	=	1000	SF
SS fins	1.65	K_{6+2}	=	1000	SF
Pour fins[d]	17.00	K_L	=	13.48	CY

[a]Cost to set MH ring and cover, cost to fit wall forms at pipes or ducts, and cost to set rungs are additional. K_{C+L+F} = hourly wage rate for one carpenter plus one laborer plus one finisher.
[b]Cost to make wall forms is additional.
[c]Cost to set and strip edges or beams is additional.
[d]Cost to hoist and/or buggy concrete is additonal.

The square foot cost for grade slabs assumes the use of a rotary finishing machine and includes the costs of setting screeds, curing the slab, and covering it with building paper on completion. The square foot price for sidewalks assumes hand finishing and includes the costs of a light trowel or broom finish as well as the costs of joining, edging, and curing.

The cubic yard cost is in addition to the square foot cost and includes the costs to place that additional concrete contained in scoop footings, thickened edges, and stud wall curbs or in slabs greater than 4 inches in thickness. The unit cost shown for forming, pouring, and finishing stud wall curbs represents an additional cost for this work, over and above the cubic yard cost discussed in the preceding.

All grade slab concrete is carried in a single column on the takeoff sheet, and there will be separate columns for the slab areas, edge forms, stud wall curbs, and hand excavation. Specification fill, vapor barrier, and mesh are carried in the miscellaneous column. The volume of concrete in excess of 12.3 CY per MSF of slab area is noted below the total for the column containing the slab areas.

When specification fill is required, the material is purchased by the ton, and its compacted density will range from 1.5 to 1.7 tons per CY. Two fine grades must be made, one for the fill and a second one for the slab.

The unit costs shown for fine grading assume essentially flat or uniformly sloping slabs and sandy loam soil. When many grade breaks occur, these unit costs may increase by as much as 60 percent. All fine grading costs assume the machine grade to be within ± 0.1 foot of finish grade.

The unit costs shown for sidewalks and curbs are typical for cases where the general contractor does the work with his own forces. When appreciable amounts of such exterior concrete are involved, savings can usually be achieved by having this work done by subcontractors.

Note: Labor and material costs as well as unit costs used are for illustration only and do not reflect present-day costs.

Cross Reference: Wiley Series of Practical Construction Guides, *Construction of Prestressed Concrete Structures* by Ben C. Gerwick, Jr. and *Construction with Reinforced Concrete* by Peck and Konon.

Masonry

An accurate takeoff is most important in estimating masonry construction. Separate consideration will be given to estimating masonry partitions and exterior masonry.

9.1 INTERIOR PARTITIONS

Takeoff of masonry partitions is not simply taking off the linear feet of partitions and multiplying the height of the wall. On most plans there are several types of wall finishes and types of block or bricks used. Each must be taken off separately and priced accordingly. Unit cost must be developed for each type of block, brick, and wall finish required. Several types of blocks used in construction today are concrete block or lightweight block made with lightweight aggregate which may be specified as load bearing or non-load-bearing with varying specifications of strengths and hourly fire ratings. When installed, structural glazed tile block known commonly as PG block or glazed concrete block require no additional wall treatment. Because of their finished surfaces, they require additional time in handling and setting. Sizes may vary. Common concrete blocks or lightweight blocks usually are $7\frac{5}{8} \times 17\frac{5}{8}$ or $7\frac{5}{8} \times 15\frac{5}{8}$ by their thickness. Glazed concrete block will usually come in the same sizes. Structural glazed tile blocks (PG) come in a variety of sizes, and the estimator must be aware of the sizes used because of the finished faces of the glazed concrete block and the PG block and added takeoff is required for special shapes required in the construction at corners, partition ends, window treatment, and so on.

Most plans will designate types of wall in plan by symbols or designate wall type by number referring to typical sections so designated on drawings of typical wall sections. The typical sections may also indicate if the walls extend to the arch or just past the ceiling elevation. One may have to refer to the ceiling plan or the finish schedule for ceiling height. The floor heights are easily discernable from the elevation or building sections.

Once the square foot areas of the different wall sections are determined, the total is multiplied by the variable unit costs developed for each section. The following is an example of determining unit cost for a typical 8-inch masonry wall struck up two sides. The labor rates used will have been determined as illustrated in Chapter 2, with all costs included except overhead and profit. Costs involved consist of the bricklayer, the mason tender, material used, plus overhead (OH) and profit. It is also assumed that for each four bricklayers, three mason tenders are required.

Assume a bricklayer can lay 120 block in a 7-hour day:

Bricklayer	7 hours at $36 per hour	252.00
Mason tender	7 hours at $33 per hour × .75	173.25
Cost of labor for each block		425.25/120 = 3.544
Block 1.20 each + waste (0.10 × 1.20)		1.20 + 0.12 = 1.320
Mortar and reinforcement (estimated)		= 0.20

Mixer, masonry saw, and scaffolding in overhead; if walls are more than 10 feet high, additional costs may be involved for scaffolding

	5.064
+ 10% OH	0.506
	5.570
+ 10% profit	0.570
Unit cost per block	6.040

The 10 percent profit margin is common for large jobs, but for medium and small jobs the profit may well warrant 20 to 30 percent markup.

After determining the total cost of the job's labor and material, there may be extra costs involved in handling and loading the job. Hoist time, difficulties in unloading trucks, or special handling equipment may be required. These costs must be evaluated and added to come up with the overall cost of the job.

9.2 EXTERIOR WALL

Exterior wall can be composed of several types of masonry configurations. The most common are brick veneer where the backup is wood or metal studs covered with plywood or other building board material. The masonry cavity wall is quite extensively used, consisting of a brick facing with a 1- to 2-inch cavity that is empty or filled with insulation followed by a block backup tied together with truss-type reinforcing such as "durawall" or special wall ties. Solid brick walls with two or more layers of brick are seldom used today. Types of block masonry wall are now used for all types of buildings taking advantage of several architecturally designed block.

Brick sizes now vary, and the size of the brick must be determined from the specification. Since takeoff is in square feet, the number of bricks per square foot must be determined. As an example, a nominal size brick $2\frac{3}{8} \times 7\frac{5}{8}$ will require 7 bricks per SF.

When taking off exterior masonry, exterior elevation drawings in conjunction with floor plans and wall sections as reference are used. Most buildings are not a simple square or rectangle. Where there is backup block, the square foot area of the backup material is less than the face material. In estimating the cost of the masonry, window areas are not deductible except if they cover a major area of a wall. Not deducting window areas compensates for waste and the added cost involved in building the jambs. When ordering material, once one has the job, the window areas are taken into consideration. The face brick usually extends below the top of the foundation wall, with the brick shelves formed in the wall to receive the brick. See the example in Figure 9.1.

The elevation of the brick shelf may be shown on the architectural drawing, but in most cases one must go to the foundation drawing for the elevation and length of the brick shelves. Where neither drawing shows the location, one has to go to the plot plan showing the finished grades around the perimeter of the building and estimate the approximate elevations of the brick shelves.

Included in the masonry takeoff are stone or precast window sills and copings. The specifications may or may not require them to be supplied by the masonry

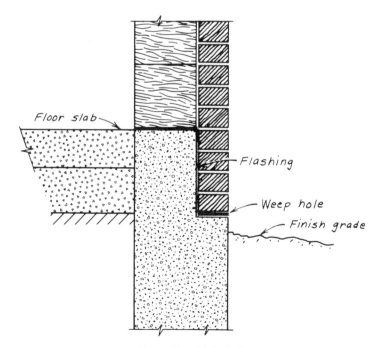

Figure 9.1. Brick shelf.

contractor, but in the case of precast, the mason will usually be required to install them when purchased by others.

The cost of scaffolding is a direct cost for the masonry contractor to figure. Whether it is pipe scaffold or hanging scaffold, the cost is a major item in the erection of the exterior masonry wall.

The cost of hoisting material is another fixed item that has to be included. The cost of the operations engineer for the hoisting equipment must also be taken into account.

A unit cost must thus be developed for the brick work and the masonry backup if required. An example of determining the unit cost of the brick work is as follows: Assume 7 bricks per SF, 350 bricks per day per mason, and $450 per 1000 bricks. With each mason add three-fourths of the cost for a mason tender. Assume the cost of a brick layer for 1 day with all taxes, benefits, and insurance is $252 and that of a mason tender is $231 \times 0.75 = \$173.25$

Brick layer	$252.00	
Mason tender	$173.25	
Cost for labor each brick	$425.25/350	$1.215
Brick		$0.450
Sand and cement, reinforcement (estimated), 12.005		$0.05
		$1.715 each brick

Thus the cost per square feet is $7 \times 1.715 = \$12$ per square foot.

The basic cost of a building's masonry cost is as follows:

1. Exterior unit cost of brick times quantity.
2. Exterior unit cost of block times quantity.
3. Interior unit cost of block times quantity.
4. Scaffolding cost.
5. Hoisting equipment cost.
6. Operating Engineers cost.
7. Winter protection and heat for mortor costs, if necessary.
8. Shanties cost.

—————————

Total cost 1 to 8

1 to 8 times percentage for overhead	_____
Total cost plus overhead	_____
Total cost plus overhead times percentage of profit	_____
Total cost plus overhead and profit	_____

CHAPTER 10

Structural and Miscellaneous Steel

10.1 STRUCTURAL STEEL

Estimating structural steel requires the takeoff of each member shown on the floor plan, column schedule, detail sections, and elevations for girts and lintels. The weights of beams are designated; for example, for a 10-inch-deep BEAM, the designation may be 10 WF 24, indicating the weight per foot is 24 pounds. Angles are designated on the drawings as for example, L3 $\times$ 3 $\times$ $\frac{1}{4}$, not indicating the weight. It is therefore necessary to go to the American Institute of Steel Construction (AISC) handbook or a supplier's handbook for the weights. Plates are designated by size and thickness. Weights must be obtained from the AISC or supplier's handbook and are designated in pounds per inch of thickness per square foot. The weights of anchor bolts and other miscellaneous items requiring takeoff will also be found in the AISC handbook since, in most instances, connecting angles with their bolts are not detailed. The weights are commonly estimated as 10 percent of the total tonnage taken off. Thus, the total tonnage of the project is the sum total of all the Bms, angles, plates, and bolts detailed on the engineer's drawing plus 10 percent allowable for connections.

Since fabrication and erection vary a great deal with the job and location as well as fabricating shops in different sections of the country, average unit prices per ton have been developed by several cost-estimating books. These books will give the general contractor a basis for budget estimating his job. The prices will be broken down into material and erection costs. The contractor may solicit cost from fabricators and erectors on a unit price basis or on a total cost basis. Having solicited a price from his subcontractors, there are certain exceptions the subcontractors will take, such as permits, site preparation for their equipment, and temporary protection, which must be added to the fabricator and erector's total cost for the supply and erection of the structural steel. The cost of shop drawings are usually included in the fabricator's price.

Steel decking is part of the structural steel supply. The decking is taken off on a square foot basis. When requesting prices from a supplier, he will adjust for splices and waste, giving the general contractor a unit price which takes into account splices and waste. Included will also be shop drawings showing splices and layout. If concrete stops are required, these must be added to the takeoff, and a separate unit price is required. The erection unit price may be supplied by the structural steel fabricator or a specialty erector engaged in erecting metal decking. Metal panels for the exterior of buildings are also taken off on a square foot basis. Girts and special attachments may be supplied by the supplier and erector of the metal panels if they are not detailed on the engineer's drawings. If detailed, the girts will be included in the structural steel cost and supplied and erected by the structural steel subcontractor.

Prefabricated steel buildings are designed and supplied by a number of prefabricated building manufacturers. The general contractor may only be required to supply the foundation, and the building is supplied and erected by the representative of the prefabricated building manufacturer. For estimating purposes unit price per square foot data is available for several types of prefabricated steel buildings through the building manufacturer's representatives.

Steel joists are taken off by weight per foot. For example, a 24-inch-deep joist will be designated as 24 J 54, which indicates it is 24 inches deep and weighs 54 pounds per linear foot. The unit price will be based on tonnage for both supply and erection. Steel joists are usually bought direct from the manufacturer or their representative and erected by the structural steel subcontractor or a specialty subcontractor engaged only in this phase of the work.

10.2 MISCELLANEOUS STEEL

Steel stairs, loose lintels, ballasts, railings, curb angles, equipment supports, miscellaneous anchor bolts, and so on, are some of the many items covered by this section of the specification. Each has its individual method of takeoff.

Steel stairs are usually taken off by the number of risers and width and type, such as pan treads, open risers, and so on, for pricing by the fabricator. The stair railings are included in the price if they are steel pipe railings. Any other type must be taken off separately. Railings are taken off per linear foot with indication of the number of horizontal rails. Lintels are taken off by weight as are curb angles and equipment supports. Anchor bolts and pipe ballasts are taken off by diameter and length. The miscellaneous steel section of the specification may include many items of equipment, which will be taken off for pricing by suppliers. Catalog numbers describing the equipment, with sizes, length, or number required, will be required for obtaining prices from suppliers. The quantity needed will sometimes dictate the unit price.

For both structural steel and miscellaneous iron, it is also important to know the requirements of the local sales tax laws and to add the necessary tax to the costs of the item.

Cross Reference: Wiley Series of Practical Construction Guides, *Construction of Structural Steel Building Frames,* 2nd ed., by William G. Rapp.

Roofing, Sheet Metal, Waterproofing, and Caulking

Estimating roofing, sheet metal, and waterproofing requires the estimator to take off the square foot areas of the roofing areas, linear foot of the flashing, and accessories required to make a watertight roof.

Caulking is taken off per linear foot, which will include any backup material.

11.1 ROOFING

There are several types of roofing systems, but basically they all require similar breakdowns in estimating costs. Following is a sample of a typical flat roof takeoff.

The square foot area of the roof is first taken off. Figure 11.1 indicates the treatment at the parapet walls. The linear feet of cant strips are taken off as well as the width and linear feet of the cap flash, which is usually copper or stainless steel. Flashing up the wall may be the same metal as the cap flashing or other suitable flashing material and is taken off by length and width.

There may be several penetrations through the roof, such as vent pipes, requiring sleeves and counter flashing, as illustrated in Figure 11.2. Pitch pockets may be required, as illustrated in Figure 11.3. Special flashing around equipment curbs and skylights must all be taken off separately and priced. Hoisting equipment must be figured separately and included in the total cost.

Roofing is usually figured per "square," which is 100 SF, thus roofing material prices are given in squares. The following is a typical checklist of items one must include in estimating a flat roofing job. The asphalt or pitch is figured in tons.

1. Roofing material: square foot area of roof times number of plies plus percentage added for waste and laps.
2. Asphalt or pitch: tonnage recommended by roofing association per square of roofing.

Metal through wall cap flashing

Metal or fabric flashing

Built-up roofing

Cant

Roof insulation

Wood blocking

Figure 11.1. Flashing at parapet wall.

Vent pipe

Pipe sleeve

Built-up roofing

Wood blocking

Roof insulation

Figure 11.2. Vent pipe sleeve.

Figure 11.3. Pitch pocket.

3. Labor installing roofing material.
4. Insulation material: square foot area of roof plus ±5 percent added for waste.
5. Labor installing insulation.
6. Cap and counter flashing material per square foot.
7. Labor installing cap and counter flashing.
8. Counting sleeves for unit price.
9. Counting pitch pockets for unit price.
10. Labor installing sleeves and pitch pockets.
11. Flashing around equipment curbs and skylights—material—square foot.
12. Labor installing around equipment curbs and skylight flashing.
13. Cants and wood blocking material per linear foot.
14. Labor installing cants and wood blocking. (*Note:* Wood blocking is usually supplied and installed by the carpentry subcontractor.)
15. Hoisting costs.

11.2 WATERPROOFING AND DAMPPROOFING

There are several types of waterproofing most commonly encountered. Membrane waterproofing applied to the outside of foundation walls and under slabs on grade for basement or pump houses below the water table. For the waterproofing of the slab, a work slab must be poured and the waterproofing membrane applied on top of the work slab. The floor slab is then poured on top of it. The membrane waterproofing under the floor slab is turned up the wall to lap with the wall membrane waterproofing, creating a waterproof envelope around the subsurface structure. Only the waterproofing material and installation will be figured by the waterproofing subcontractor; the work slab will be constructed by others. Protection of the waterproofing from the backfill will usually be a 1- to 2-inch protection board stuck to the wall waterproofing. Protection of the waterproofing on the work slab may be required if there will be an unusual amount of activity over it in preparation of the floor slab pour.

The following is a checklist for the preparation of a membrane waterproofing cost estimate:

1. Membrane material per square foot.
2. Asphalt material per ton.
3. Protection board material per square foot.
4. Labor installing membrane.
5. Labor installing protection board.
6. Scaffolding if wall is high and backfill does not keep up with the installation of the waterproofing.

Waterproofing elevator pit walls are usually carried out on the inside face of the pit. Ionite waterproofing is most commonly applied $\frac{3}{4}$ to 1 inch in thickness. It is troweled on the walls and slab of the pit. The pit wall must be roughed up before the Ionite is applied. Besides the labor and material of the waterproofing, the cost of the application of this type of waterproofing must include the expensive cost of preparation.

Waterproofing of exterior brick and block surfaces is usually applied with a liquid waterproofing material such as silicon. Manufacturer's recommendations on the material used gives one the approximate area covered by the material. Using these estimates of cover, the estimator has a guide in figuring the amount and cost of the material. The labor costs can thus be judged by the amount one is able to apply in an hour or a day. If no scaffold is provided by others, the cost of the scaffolding must be included.

11.3 CAULKING

As stated at the beginning of the chapter, caulking is taken off by linear feet. The estimator must take off the total linear feet of the applied surfaces and note the thicknesses of the joints. Manufacturers recommendations of the material used for each application will give the estimator a good idea of the cost of material. At least 10 percent should be added for waste and backup rod material where they are required are easily figured. Many times the installer may not have a full day's work, but a full day must be figured if such a situation occurs. Normally, scaffolding is provided by others. Where the applicator must provide his own scaffolding, the additional cost of providing scaffolding must be included in his overall cost of the job

Cross Reference: Wiley Series of Practical Construction Guides, *Construction Sealants and Adhesives,* 2nd ed., J. R. Ponek and J. P. Cook, P. E.

CHAPTER 12

Carpentry

There are two broad categories of carpentry work, rough carpentry and finish carpentry. For estimating purposes, these two types of work are taken off separately but are priced in a single section of the job estimate.

The format for the carpentry estimate sheet is identical to the one described at the beginning of Chapter 8 for structure concrete, and the procedures followed to estimate the time required for completion are identical to those illustrated in the sample estimates in Chapter 8.

Rough and finish carpentry are usually covered in two separate sections of the specifications. In addition to the work encompassed by these sections, the installation costs for certain materials covered under other sections of the specifications are included in the carpentry estimate.

Thus hollow metal doors and frames, millwork, and finish hardware are all furnished, FOB the job, under separate sections of the specifications. The costs of these materials are entered in the appropriate sections of the estimate summary sheet, but the costs for their installation are carried in the finish carpentry estimate. The suppliers of these materials make their own takeoffs in all cases, tender their quotations on bid day, and are, in this respect, similar to subcontractors.

The lump sum quotation for millwork normally includes the costs of all wood doors, unfinished wood cabinets, milled trim pieces, and that rough lumber which requires milling and which is detailed in the job plans. In addition, the millwork quotation may include a square foot cost for hardwood finish plywood.

Prefinished wood cabinets, when specified, are covered under a separate section of the specifications. They are furnished and installed, complete, by the manufacturers on a subcontract basis. Prefinished wood cabinets are furnished with all doors installed, and no additional finish hardware, painting, or work by the general contractor is required.

319

The lump sum quotation for finish hardware will include the costs of all hardware required for metal doors, wood doors, and unfinished wood cabinets and all accessories such as door closers, door stops, and metal thresholds.

Toilet stalls are covered in a separate section of the specifications and are usually furnished and installed by specialty subcontractors. When quoted FOB job, toilet stalls are installed under finish carpentry. In all cases toilet stalls are taken off by the suppliers.

12.1 GENERAL COSTS

The general costs section for carpentry work will include the following items:

1. Lumber.
2. Plywood.
3. Miscellaneous iron.
4. Other materials.
5. Nails.
6. Material suppliers or subcontractors.
7. Hand power tools.
8. Hand tools.
9. Saw filing.
10. Compressor.
11. Power supply.
12. Table saw.
13. Hoisting.
14. Other items.

The rental periods used for equipment items on the preceding list cover only the time the equipment is required for carpentry work. Care must be exercised to avoid overlapping the rental periods for those items previously included in the general costs section for structure concrete, which may be available, as needed, for the carpentry work.

12.1.1 Lumber and Plywood

Lumber is, in general, taken off piece by piece, and a combined lumber list is prepared and distributed to the lumber yards for materials quotations. In all cases the lengths are listed in multiples of 2 feet. Thus, five hundred 2 × studs 8 feet 6 inches long would be listed as 2 × 6, 500/10. When these studs are spaced 16 inches on center, the waste will furnish one piece of fire blocking, and no additional allowance for waste would be necessary in connection with these two items.

Where materials such as ledgers or stud wall plates are taken off by the linear foot, an allowance for waste must be made, and its size will depend on the plan dimensions and on the estimator's judgment. Thus, if the average length of run were slightly greater than 30 feet, the estimator would plan to use two pieces 16 feet long, and the factor for waste would be approximately 32/30 = 1.07. In such a case, the neat linear feet taken off would be increased by 7 percent, and the resulting figure converted to an even number of pieces 16 feet long.

Where boards are used as sheathing or as decking, the finished board widths must also be considered when determining waste allowances.

For example, assume that a long section of stud wall having an overall height of 9 feet is to be covered with 1×6 diagonal sheathing. The sheathing is to be laid at 45 degrees, and the finished board widths are $5\frac{1}{2}$ inches. The diagonal length will be $1.414 \times 9.00 = 12.73$ feet, and the required board length will be 12.73 feet plus $5\frac{1}{2}$ inches $= 13.19$ feet.

The boards will be ordered in 14-foot lengths, and the waste factor applicable to the gross sheathing area will be, in this case, $14/13.19 \times 6/5.5 = 1.16$. A waste allowance of 20 percent is often used for diagonal sheathing.

Whenever possible, the plywood takeoff is based on the actual number of sheets required. Thus, a roof area measuring 31×102 feet is considered as 32×104 $= 3.328$ MSF and would require 104 sheets 4×8 feet in size.

On all jobs, there will be some sections where the allowances for waste cannot readily be determined. In such cases allowances of 10 percent for waste lumber (due to end cutting) and 15 to 20 percent for waste plywood have been used.

12.1.2 Miscellaneous Iron

This item covers the purchase price, FOB job, of the bolts, joist hangers, and other miscellaneous iron required for the carpentry section. The cost to set those bolts which are embedded in concrete is included in the operating section of the carpentry estimate relating to the lumber requiring the bolts. The costs to install bolts connecting wood to wood are included in the unit costs for the various types of carpentry work.

12.1.3 Nails

Nailing schedules for the major types of work are given in the plans and specifications. The theoretical number of nails required for a given type of work is estimated by considering a typical piece or section and extrapolating this figure based on the ratio of the total board feet involved to the board feet in the piece or section considered. The theoretical amount of nails so determined should be increased by about 20 percent to allow for waste, loss, and pilferage.

As a rough check, approximately 100 pounds of nails are required per 4 MFBM or per MSF of lumber or plywood involved, and no additional allowance for waste is necessary.

12.1.4 Material Suppliers or Subcontractors

Timber trusses and laminated wood beams are often included in the carpentry section of the estimate. These materials are usually quoted FOB the job by prefabricated timber products suppliers.

In such cases, the estimator should set up separate sections on the estimate summary sheet to cover the purchase price for these materials. This procedure ensures that the quotations for these materials are not omitted from the estimate during the bid day rush and permits the estimate sheet for the carpentry work to be completed and checked prior to bid day. The installation costs for trusses and laminated wood beams are carried in the estimate for the carpentry work. Lamella roofs are normally furnished and installed by timber engineering contractors on a subcontract basis and are usually covered by a separate section of the specifications. When lamella roofs are included in the carpentry section of the specifications, the estimator should set up a separate section for this work on the estimate summary sheet.

12.1.5 Hoisting

The rental charges for the truck cranes required to hoist or to erect carpentry materials are carried in the operating sections of the carpentry estimate. The item is included on the checklist to ensure that the estimator does not overlook this cost when preparing the estimate.

12.2 UNIT COSTS

In general, the unit costs derived for rough carpentry work exhibit greater variations from job to job, from day to day, or from one crew to another than do those for any other type of work performed by the general contractor. Subtle differences in the framing details, in the amounts of work that can be done at any one time, and in the quality of the foremen, the crews, and the job superintendent can result in cost variations as great as 100 percent.

When framing subcontractors are available, savings can usually be achieved by having them do this work. They are rough carpentry specialists, they do this work day after day, and their crews are highly skilled.

Unit costs for rough carpentry work are shown in Table 12.1. These costs represent average values for first-class public work and assume that the quantities of work that can be done at a given time are large enough to enable the crews to establish repetitive daily patterns and the rough carpentry can be prosecuted continuously for a period of several weeks. These unit costs include the costs for the saw men (table saw operators) and flat rack truck drivers if and when required.

In all cases, the unit costs for carpentry work are applied to the total MFBM or MFS of material purchased, including the allowances for waste.

Unit costs for finish carpentry work are shown in Table 12.2, and these costs are applicable to first-class public work performed by the general contractor's own forces.

TABLE 12.1 Unit Costs for Rough Carpentry

Activity	Cost of Labor ($), $K_{6+2} = 1000$	Unit of Measure
Stud walls, H8-H10 (includes sills, plates, studs and blocking:		
2 × 4	590	MFBM
2 × 6	520	MFBM
Additional		
Set anchor bolts for sills	1.20	Each
Dry pack sills and set to grade	2.20	LF
Floor joists (includes blocking):		
2 × 8	430	MFBM
2 × 12	400	MFBM
Roof joists (includes blocking):		
2 × 8	450	MFBM
2 × 12	430	MFBM
Ledgers (bolted to concrete walls):		
3 × 8	700	MFBM
3 × 12	650	MFBM
Set ledger bolts in wall forms	1.50	Each
Furred Spaces (2 × 3 material)	800	MFBM
2 × 3 Stripping (to ceiling joists)	650	MFBM
2 × 4 Subfloor sleepers (metal clips additional)	760	MFBM
2 × 6 Framed platforms	720	MFBM
Fascia boards:		
2 × 8	1000	MFBM
2 × 12	860	MFBM
Eave framing (2 × 4, 2 × 6)	660	MFBM
Purlins:		
2 × 3 bolted to steel @ 4' centers includes drilling 3/16″ steel	720	MFBM
2 × 4 screwed to steel @ 4' centers	430	MFBM
2 × 6 on wood (blocking additional)	400	MFBM
2 × 6 T & G roof sheathing	250	MFBM
Plywood (blocking additional):		
Subfloors 3/8″	260	MSF
3/4″	280	MSF
1 1/8″	350	MSF
Shear walls 3/8″	400	MSF
Roofs 1/2″	220	MSF
3/4″	280	MSF
Exterior walls (3/8″ shingle panels 8'0″ × 16″)	750	MSF
4 × 2 Blocking for plywood	790	MFBM
1 × 6 Diagonal sheathing (walls)	500	MSF
Battens on wood walls:	0.60	LF
1 × 4		
2 × 4	0.86	LF
Plaster grounds, 3/4 × 2	0.60	LF

TABLE 12.2 Unit Costs for Finish Carpentry

Activity	Cost of Labor ($), $K_{6+2} = 1000$	Unit of Measure
Doors		
Set hollow metal frames (to concrete)	49	Each
Hang hollow metal doors complete	21	Each
Set and trim wood jambs	36	Each
Set rough wood bucks	14	Each
Hang wood doors complete	44	Each
Block out, set, and grout Rixson-type hinges	63	Each
Set frame and hang vault door (32′ × 78′)	160	Each
Set frame and hang refrigerator door	170	Each
Install panic hardware (additional)		
Wood jambs, per panic bar	42	Each
Hollow metal frames, per panic bar	63	Each
Install door closers	22	Each
Install door stops		
Wall mounted	5.50	Each
Floor mounted (concrete)	10.00	Each
Install metal thresholds	18.00	Each
Install name plates	1.10	Each
Install kick or push plates	6.30	Each
Install weather stripping (per door)	11.00	Each
Install unfinished wood cabinets (per SF of face area)	1.20	SF
Open shelves		
Typical cabinet installation	1.55	SF
In congested areas	3.00	SF
Install cabinet doors (additional)		
Small	15	Each
Large	20	Each
Install interior finish plywood		
¼″ DF walls	470	MSF
¼″ DF ceilings	870	MSF
¼″ Birch walls	890	MSF
Install toilet stalls		
Floor mounted	46	Each
Ceiling hung	96	Each
Install shoe mold or wood base	0.60	LF
Install glass stop (¾″ × ¾″)	0.54	LF
Install alluminum letters on concrete H6″–H8″	11.50	Each
Install bronze plaques on concrete	10.00	SF
Aluminum sash (glazing additional)		
Set	1.10	SF
Caulk ¼″ × ¼″—with gun)	0.25	LF
Grout regelets	90[a]	CF
Roll-up doors (manually operated)[b]		
Set aluminum doors	5.50	SF
Set steel doors	4.00	SF

[a]Based on $K_r = 16.10$.
[b]The equipment cost for handling the doors is additional.

The unit costs shown for installing unfinished wood cabinets are based on the areas of the front faces of the cabinets, and a single shelf is assumed to have a frontal area of 1 SF per LF of shelf.

Experienced estimators often assign a lump sum cost per cabinet for installation, and the lump sum used is based on judgment and a careful consideration of each cabinet to be installed. In all cases, the cabinet sizes must be considered with respect to the room sizes and the doorway widths, and it may be necessary for some of the cabinets to be furnished in sections and assembled in place.

12.3 HEAVY TIMBER BENTS

The costs to frame and to erect two similar pieces of timber having identical cross sections will be essentially the same even though their lengths may vary by 100 percent or more. Thus, the costs for heavy timber work cannot be based on the board feet involved except in those cases where cost records from virtually identical work are available.

Structures such as heavy timber bents are framed and assembled, insofar as possible, on a level work area or platform which may consist of two or more rows of 10×10 or 12×12 timbers blocked to grade. The platform should be as close to the point of erection as possible, and the material to be framed should be delivered nearby.

The timbers are placed on the platform with light equipment, such as an A-frame truck or a small rubber-tired backhoe, and are positioned and framed by carpenters using peaveys and cross-cut saws. The assembled sections are then erected with truck cranes.

Two carpenters working on a platform will position, set straight edges, and make one square cut on a 12×12 timber in about 0.33 hours, and this time should be increased by 50 percent if two cuts are required. About two-thirds of the 0.33 hours will be spent handling and positioning the material, and this time will be essentially the same for timbers larger than 12×12. Thus, the estimated time required to make one square cut and to position a 16×16 timber would be about $\frac{2}{3} \times 0.33 + \frac{1}{3} \times 0.33 \times \frac{256}{144} = 0.42$ hours for two carpenters.

In many cases some additional time is required for a laborer to furnish blocking and to assist the carpenters. The amount of laborer time allowed depends on the estimator's judgment Indexed to $K_{6+2} = 1000$, a two-carpenter crew will cost 1000 $\div$ 24 24 $= \$41.67$ per hour; this figure includes $\frac{2}{3}$ hour of laborer time.

Two carpenters can make one square cut on an 8×8 timber in about 0.11 hours, and about one-half of this time is spent handling the material.

One carpenter can lay out and drill one hole through a 12×12 to 16×16 timber in about 0.08 hours and install one bolt with washers in about the same amount of time.

Based on $K_{6+2} = 1000$, the cost to install 3×8 bolted bracing, complete, on horizontal bents is about \$350 per MFBM and the cost to install 3×8 bolted bracing on erected bents is about \$500 per MFBM.

The equipment rental costs to handle and to erect the bents depend on the size of the assembled units, the equipment that is available, and the number of assembled units that can be erected at one time.

Figure 12.1 illustrates a cost estimate for fabricating, erecting, and dismantling two timber falsework bents. The labor costs are indexed to $K_{6+2} = 1000$. It is assumed the fabricating platform is located near the point of erection and that the materials are located close to the platform. The unit costs used are derived as follows:

Line 1. Each platform timber will be blocked to grade at two locations. The unit cost to SS flat pads shown in Table 8.10 is used. In this case the pad loading will be small, but each pad must be blocked to a level grade.

Lines 2 and 3. We assume a two-carpenter crew plus a small rubber-tired backhoe will set 16 pieces of timber on the pads or on the platform per hour. The bent timbers will be positioned by the carpenters as part of the framing operation.

Line 4. We assume the backhoe is located on the job. The second bent will be fabricated on top of the first one. Two on-and-off charges are required, one to set the timbers for the platform and the first bent and another to set the timbers for the second bent.

Line 5. Two cuts are required for each post, one to square the bottom and another at the top to accommodate an assumed cross-fall at the cap. Thus, the time required per post will be about $(0.33 \times \frac{2}{3} + 0.33 \times \frac{1}{3} \times \frac{196}{144}$ $1.5 = 0.55$ hours. Assuming a two-carpenter crew, the cost per post will be $0.55 \times 41.67 = \$23$.

Line 6. The time required to position and to make one square cut on each 14×16 timber is estimated to be $0.33 \times \frac{2}{3} + 0.33 \times \frac{1}{3} \times \frac{224}{144} = 0.39$ hours. Assuming a two-carpenter crew, the cost per piece will be $0.39 \times 41.67 = \$16$.

Lines 7 and 8. $0.08 \times 41.67 \times 0.5 = \$1.70$.

Line 10. We assume a four-carpenter crew plus a truck crane will erect each bent in $\frac{1}{2}$ hour.

Line 13. We assume the posts are set on sandjacks and that one laborer releases four jacks per hour.

Line 14. After the sand jacks have been released, the falsework supported by the bents will be removed laterally, and the bents will be toppled. We allow an additional $\frac{1}{4}$ hour for a four-carpenter crew to topple each bent.

Line 15. The bents were framed on a platform but will be dismantled on the ground. This requires more time, and we therefore use the total costs for lines 7,

Figure 12.1. Cost estimate for timber bent.

8, 9, and 11. The inclusion of line 7 in this total as well as the inclusion of line 11 at full cost compensates for the extra time required to release the drift-pinned connections between the caps and the posts.

12.4 MISCELLANEOUS CARPENTRY COSTS

Galvanized steel bench supports made of $\frac{3}{8} \times 2\frac{1}{2}$-inch material were bolted to a concrete slab with three $\frac{1}{2}$-inch expansion bolts each at a unit cost for labor of $29 per bracket, indexed to $K_{6+2} = 1000$.

Eighty-five MFBM of 2×6 seat boards were installed on brackets at 4- to 6-inch centers using two $\frac{3}{8}$-inch bolts per board per bracket at a unit cost for labor of $360 per MFBM. This unit cost was achieved by a skilled and well-organized crew. Small quantities of seat boards will be installed for a unit labor cost of about $420 per MFBM.

Roof vents averaging 30 inches square were framed, complete, at a unit cost for labor of $23 each, indexed to $K_{6+2} = 1000$.

Temporary job offices and sheds made of prefabricated panels will be erected and removed at unit labor costs ranging from about $4 to $6 per square foot of floor area, indexed to $K_{6+2} = 1000$. The higher unit cost applies when interior doors and partitions are required.

Rough 3×8 caps 6 feet long were bolted to steel shoring frames with eight $\frac{3}{8} \times 2\frac{1}{2}$-inch lag screws at a unit cost for labor of $2.90 each, or $242 per MFBM, indexed to $K_{6+2} = 1000$.

Note: Labor, material, and unit costs used are for illustration only and do not reflect present-day costs.

Cross Reference: Wiley Series of Practical Construction Guides, *Construction with Timber* by R. Kaseguma.

Hardware and Hollow Metal

13.1 HARDWARE

Hardware is almost always bought by the general contractor directly from a builder's hardware supplier. A builder's hardware supplier usually represents one or two of the major hardware manufacturers. If the hardware supplier usually used by the general contractor does not carry the specified hardware, substitution can be requested from the architect. Where no substitution of manufacturer is allowed, the representative of the specified hardware must be found. Therefore, it is good practice to deal with several builder's hardware suppliers. In some areas, there may be more than one builder's hardware representative of a particular manufacturer, and there is room for price negotiation.

Hardware is called for on the drawings by hardware set numbers or listed on the door schedule by hardware set. When listed on a door schedule, a separate column is designated for each door hardware set. Some plans do not include a door schedule but only include typical details, and the hardware sets are designated on the plan at each door opening, as illustrated in Figure 13.1.

In most cases the hardware sets are listed in the specification, but on smaller jobs they may be listed on the drawings. If the estimator cannot get his hardware supplier to take off the quantities, a simple takeoff by the estimator is required, listing each type of hardware set and the number required.

Installation of the doors and hardware are by carpenters supplied under the carpentry section. Pricing installation requires sound judgment by the estimator as to the amount of time it takes to install each item of hardware. Each hardware set is priced for installation by assigning time to each operation, not only the time required for the installation but also the time required for sorting out the hardware and distributing doors in place. Adding the cost of sorting and distribution is usually assigned to the installation of the butts (hinges). An illustration of pricing the installation of a hardware set follows. On the hardware schedule, the architect will usually list the catalog number of the specified manufacturer of each item. If one has no knowledge of the hardware specified, the manufacturer's catalog should be requested and consulted.

H. M. Door Typical Hardware Set

butts, $1\frac{1}{2}$ pairs; time includes sorting of hardware and distribution of door	$\frac{1}{2}$ hour
Lock set with escutchen to be drilled and tapped in place (each side)	$1\frac{1}{2}$ hours
Closure, surface mounted	$\frac{1}{2}$ hour
Door bumpers, wall mounted, toggle bolts	10 minutes
Total time	2 hours 40 minutes

The cost for the installation of the door will be the hourly rate of the carpenter with all benefits and insurance × 2.67 hours plus overhead and profit.

It should also be noted that a detailed hardware schedule will be supplied by the hardware supplier in his overall bid for the supply of the hardware.

13.2 HOLLOW METAL

Hollow metal is also bought direct from the manufacturer by the general contractor. Here, again, if the manufacturer does not take off the quantities, the estimator must take off each type of door and buck by quantity for pricing by the Hollow Metal supplier. The type of door and buck may be designated in the door schedule or as illustrated in Figure 13.1.

The door buck and doors are installed by carpenters, as indicated previously. The door bucks installed in drywall construction are usually included in the drywall subcontractor's bid. Door bucks installed in masonry must be priced separately for

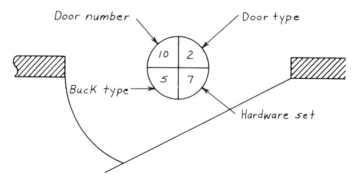

Figure 13.1. *One method of designating door openng on plan.*

installation by the carpentry contractor. Distribution cost for the door buck should be assigned to the installation cost of the door bucks. Additional time may also be required for layout if the mason excludes the location and layout of door openings in his contract.

It should be noted both for hardware and hollow metal takeoff that some door schedules indicate typical door numbers rather than openings. Door 1 on the door schedule may represent four doors. The estimator must then take off the doors and bucks from the plan, indicating the quantities of doors and bucks per typical number, for unit pricing.

Finishing Trades: Drywall, Ceilings, Flooring, and Painting

Estimating costs for the finishing trades covered in this chapter are by unit price. Therefore, the accuracy of the takeoff is most important.

14.1 DRYWALL

The estimator should not deduct for windows or doors when taking off the square foot area of the walls. Since the drywall subcontractor will include the installation of the door bucks, they must be counted for the additional cost of installation. When taking off the walls, each individual wall should be taken off separately for unit pricing.

On the plans, the walls are usually designated by detail numbers or other means of identification corresponding to typical wall section details, illustrated in the drawings (Figure 14.1). The wall sections can consist of different numbers of sheetrock layers on each face of the studs. If one has the knowledge of how many square feet of sheetrock a carpenter can install in a day or how many studs a carpenter can erect in a day, it is possible to derive a unit cost for each operation. This, of course, is available to an estimator who has the field knowledge after years of experience installing drywall. Since most estimators for general contractors have not the experience to determine such costs, they must rely on the available unit price information derived from such pricing information as ''means construction cost'' yearly pricing information. One may develop one's own prices by analysis of past contracts. Keeping records of square foot areas and actual cost of job or subcontractor prices, it is possible to develop a fairly accurate pricing structure. Once unit costs have been developed for each operation, the square foot unit cost of each type of wall can be calculated. In addition to the basic material and labor, the unit price must include accessories such as screws and waste for each operation. Overhead

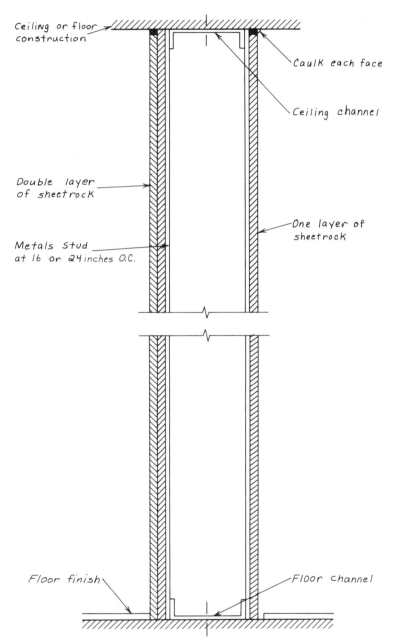

Ceiling or floor construction

Caulk each face

Ceiling channel

Double layer of sheetrock

One layer of sheetrock

Metals stud at 16 or 24 inches O.C.

Floor finish

Floor channel

Figure 14.1. Typical wall section.

and profit can be included in each unit cost or added to the final cost figure. This is an example of pricing a typical wall section:

Height of the wall = Y
First layer each side, $2 \times$ (unit price per square feet) = A
Second Layer one side, $1 \times$ (unit price per square feet) = B
Header channel, 1 foot $\times$ (unit price per foot)/Y = C
Base channel, 1 foot $\times$ (unit price per foot.)/Y = D
Studs required for 1 SF of wall $\times$ (unit price per SF) = E
Caulking at ceiling, $2 \times$ (unit price per foot)/Y = F
Taping and spackling each face $\times$ (unit price per SF) = G

Therefore, the total price of this wall detail per square foot is $A + B + C + D + E + F + G$. After the total cost of the individual walls have been calculated, additional costs are added such as loading the job, hoisting charges, shanties, and special labor costs such as a shop steward or a nonworking foreman. Additional costs include installation of door bucks and cats for installation of future equipment installation such as toilet accessories, wall railing, and wall cabinets. Additional cost for layout may be necessary if the job is complicated. The final cost will then be the overhead and profit if not included in the unit cost of the item making up the total cost of the job.

14.2 CEILINGS

Hung ceilings may be either drywall or accoustical and exposed tee or concealed spline. All must be suspended from the floor above. In new construction where the supporting structure is concrete, inserts, or tees (see Figure 14.2), are usually installed through the plywood or metal decking forms spaced 4 feet on center, each way by the lather. The hole at the bottom of the tee will be used to bolt a hanger strop, usually $1 \times \frac{1}{8}$ inches or a rod hooked through the hole supporting the ceiling carrier channels. For a direct hung ceiling, the tie wire will be suspended from the hole in the tee. In renovations, if no hangers exist, a hanger can be shot into the concrete. If the concrete is too soft, the concrete must be chipped away exposing the reinforcing wire or rods and the supporting wire or rods suspended from the slab's reinforcing. The area disturbed is then patched. With timber construction the hangers are secured to the side of the beams. One can see that the cost may vary considerably with the method of attachment to the supporting structure. With a sheetrock ceiling, a carrier structure called the "black iron" consisting of a $1\frac{1}{2}$- to 2-inch channel 4 feet on center is hung from the arch and leveled. Below, attached to the carrier channels, are the furring strips 16 to 24 inches on center, to which the sheetrock will be screwed. Some city building codes may require carrier channels (black iron) for support of an exposed tee accoustical ceiling instead of directly hanging the supporting tee by wire from the hangers in the arch. With a conceiled

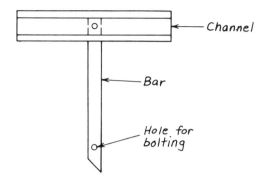

Figure 14.2. Tee hanger.

spline ceiling, black iron must be used. Most hung ceilings will have recessed light that may interfere with the black iron. Bridging between carrier channels must be installed, an additional cost to be considered. The estimator, thus, has several other factors to consider beside taking off the square foot area of the ceilings. *Note* Metal decks may now be supplied with tabs for hanging hung ceilings.

When taking off the area from the floor plan or ''reflected ceiling plan'' (a plan showing the layout of the accoustical ceiling pattern), no deduction is taken for lighting fixtures. Room sizes for sheetrock ceilings can be taken off as the actual room size. An additional 10 percent may be added for waste. When taking off room sizes for 1 × 1 tile, the room is figured to the nearest plus foot. For 2 × 2 tile the room size is figured to the nearest plus 2 feet and for 2 × 4 tile to the nearest plus feet in the 2-foot direction and 4-foot increments in the direction of the 4-foot dimension of the tile. On large floor areas, where the majority of rooms are large, it may be possible to take off the total floor area not deducting for walls.

This must be considered in pricing ceiling jobs (hangers 4 feet on center each way and carrier channels 4 feet on center; w.r. = where required):

1. Hanger material

$$\frac{\text{Floor area}}{16} + 5\% \text{ waste} = \text{number of hangers} \times (\text{straps, inserts, T-supports}) \text{ unit costs} = A.$$

2. Hanger labor

$$\text{Number of hangers} \times \text{unit costs (w.r.)} = \text{digging out} + \text{reinforcing} + \text{patching}$$
$$= B.$$

3. (w.r.) Carrier channels (black iron) labor and materials

$$\text{Length of room} \times \text{number of channels at 4 feet on center} + \text{one additional length}$$
$$\times \text{unit cost per foot} = C.$$

(*Note:* Black iron average unit cost has been developed based on the square foot area of floors, for example, $1.00 per SF.)

4. Furring material and labor

Length of room × number of channels 16 or 24 feet on center + one additional length of room × unit cost per foot = D.

5. Exposed tee material and labor

Grid and tile unit costs (based on SF area of ceiling, so much per square foot) = E.

6. Conceiled spline material and labor

Grid and tile unit costs (based on SF area of ceiling, so much per square foot) = F.

7. Sheetrock material and labor

Installation unit cost (based on SF area of ceiling) = G.

Therefore, the total cost of each type of ceiling is as follows:

Sheetrock ceiling = $A + B + C + D + G$
Exposed tee ceiling = $A + B +$ (w.r.) + E
Conceiled spline ceiling = $A + B + C + F$

Unit price per square foot of area is the general norm for estimating ceiling construction. As stated before for drywall, the estimator can develop his own unit price schedule from previous jobs or rely on published unit price such as means construction costs.

14.3 FLOORING

Wood floors, resilient flooring, and carpeting are the most common types of floor finishes. The estimator need only take off the floor square foot areas, add the necessary factor for waste, and multiply by an appropriate unit cost plus any required floor preparation. For carpeting, it may be necessary to lay out the room in order to determine the waste since carpeting comes in 9-, 12-, and 15-foot widths. The location of seams may determine the amount of waste. Resilient flooring comes in cartons of 45 SF, and with many changes in design from room to room and small rooms, the waste of unfinished cartons may have to be taken into consideration. The price of the material is always readily available, but again labor cost can only be determined through experience or relying on someone else's experience.

In addition to the cost of the flooring, additional costs may be involved in floor preparation. Concrete floors may be uneven and may have to be flash patched (smoothing and filling in of depressions), grinding high spots, and so on. Allowances are usually made for some flash patching, but if an excessive amount is encountered on the job, negotiations for some monetary relief may be in order. Where subfloors are extremely bad or if the subfloor is strip or parquet flooring, an underlayment is required. Generally, $\frac{1}{4}$-inch hardboard or $\frac{3}{8}$- to $\frac{3}{4}$-inch plywood is used.

The following is a checklist for estimating flooring:

1. Floor preparation.
2. Material cost of flooring.
3. Installation cost (labor).
4. Cleanup, such as polishing for resilient flooring.
5. Floor finishing for wood floors, sanding and sealers or polyurethane.
6. Protection coverings.

14.4 PAINTING

When estimating painting, walls and ceilings must be taken off separately. Windows, doors, and bucks must be counted since all are priced out separately. In the taking off of walls, no deduction is taken for windows or doors from the square foot areas of the walls. Doors and windows require additional time to paint. Additional labor cost (unit costs) must be added for each unit.

The application of each coat of paint varies. The prime coat may require less labor but more paint. Different unit costs may be developed for each application of paint. Additional preparation of the walls before or after each coat of paint must be reflected in the unit cost of each application. In an industry that depends so much on experience, unit cost can only be established by the knowledge of one's work force. Painting contractors do know their work force capabilities and have developed their own unit cost. Developing close relation with one's painting contractors, averaging unit costs may be solicited from them. This will enable the general contractor to come up with a fairly close budget estimate.

Preparation of the walls of old construction includes cutting out of cracks and spackling. New construction usually requires corrective work to the taping and spackling of, for example, sheetrock joints. Metal door bucks may have to be sanded down and cleaned from previous work around the door bucks. All these additional operations require consideration in the final cost of the job.

The following is a checklist for pricing the cost of painting:

1. Wall and ceiling preparation.
2. Unit cost per square foot for each coat of paint.
3. Added cost for painting doors and windows.
4. Protection of existing equipment and metal surfaces not painted.

Miscellaneous Material and Equipment

The general contractor buys many special materials and equipment items for a project which may be installed by his forces or subcontracted out for installation.

There is a wide variety of union rules governing the unloading and handling of material and equipment. Mechanical equipment that will require the services of a steam fitter, plumber, sheet metal, and electrician may require a representative from each of the trades to participate in the unloading and positioning of the equipment as well as the rigger who may be doing the unloading. The estimator must be aware of current union regulations when purchasing equipment that will require, for installation, the services of several trades. Besides the labor costs involved in the unloading and installation of the equipment, there may be transportation costs, permits for cranes unloading the equipment, extra rubbish removal costs for crates, and packaging material. Most of these costs must be considered before deciding to buy equipment direct rather than having a subcontractor responsible for the supply and installation of the equipment.

As a guide, here is a checklist of costs to consider when purchasing special materials and equipment:

1. Cost of equipment and material (subcontractors may get a better discount on purchases).
2. Transportation costs.
3. Unloading and handling costs (labor and unloading equipment).
4. Permits.
5. Trash removal (container or trucking costs).
6. Installation costs (labor).
7. Testing of equipment costs.
8. The cost of services of the manufacturer's representative, if required, during the installation of the equipment.
9. Sales taxes, where required.

Mechanical Trades and Electrical

General contractors seldom take it upon themselves to do the mechanical trades or the electrical work with their own forces. Exterior drainage and some auxiliary electrical work in conjunction with the site work may be constructed, in part, by the general contractor's own forces. Almost without exception the mechanical and electrical work are subcontracted out. During the bidding process, the general contractor must have an idea of the cost, early on in the bidding, for advising the bonding company as to the overall cost of the job or for one's own benefit. The following will give the estimator a basis for a budget estimate and/or a means for negotiating with the subcontractors as they submit their bids.

16.1 PLUMBING

One of the first steps in new construction is to supply water to the site. Costs involved are street openings, permits, tapping of water main, piping to the building area, excavation and backfill, and patching of existing street paving. Permit costs can be obtained from the municipal agency involved. Unit prices for excavation locally and installation of pipe may be available from the knowledge of past performance or published cost figures such as "means." The tapping into the main is usually accomplished by the water company employees, and the cost is obtained from the water company.

The sanitary sewer line may be connected into an existing sewer system, or if none exists, septic tanks and dry wells must be built. Permits for street openings, where necessary, as well as patching of streets are additional costs. Depending on the size of the existing sewer line and the new sewer line, a manhole may have to be built at the intersection. Where septic tanks and dry well precast units are used, costs are easily obtained from the manufacturers of the precast units.

Excavation and backfill length of pipe runs and manhole covers for manholes, septic tanks, and dry wells must be figured. Thus, the following cost must be considered for bringing in water and sewer service to a project.

1. Permits.
2. Tapping water main.
3. Street patching.
4. Excavation for piping.
5. Excavation for piping and drainage structures.
6. Labor and material costs of piping and installation.
7. Manhole covers.
8. Precast structures.
9. Installation of precast structures.
10. Special backfill such as gravel or crushed stone around the precast structure.
11. Dewatering, if required.
12. Sheathing of excavation, if required.

Plumbing in the interior of a building is usually estimated based on a fixture unit price—so much for a sink, so much for a bathtub, and so on. The count and a unit price will give the general contractor estimator a fairly good idea of the interior plumbing costs. The unit price of the fixture will include the roughing as well as the fixture, labor, and material costs. Thus for a sink which costs $150, the unit price for the sink in place may be $300. This will hold true in multistory buildings such as offices and apartment buildings. In a warehouse or factory where only a few toilets are involved, the unit price of the fixtures can still be used, but depending upon the location, it may be necessary to add the additional cost of pipes leading from the incoming service to the toilet area. A floor drain system for the warehouse or factory building must be estimated by taking off the lengths of pipe, excavation, if involved, and the cost of drains and accessory structures.

Roof drains can be estimated on fixture unit bases if they are tied into the sanitary lines servicing the toilets. If a separate line is used for the drainage system, the risers, cleanouts, and extra runs to the sewer or dry well system must be priced up separately.

Standpipe systems with fire hoses and cabinets can be easily estimated. The supply line unit prices are easily available, and hoses and cabinet prices can be obtained from suppliers. Usually a separate meter system for the standpipe and sprinkler system is supplied under the plumbing specifications. Cost for the meter can be obtained from a plumbing supply house. Installation cost can easily be estimated. A foundation pad may be needed in some instances.

A checklist of costs for interior plumbing is:

1. Office and apartment buildings: Count the number of each type of fixture assigning a unit price to each which will include all labor, material, and roughing required for the fixture installation.
2. One-story warehouse or factory:

    a. Drainage system: Length of pipes, excavation, if necessary, and number of drains.

    b. Toilet area. Fixture count for unit pricing. Take off of length of water and soil piping to toilet area for unit pricing. Excavation costs, if necessary.

3. Boiler and controls for hot water.

4. Standpipe system: Counting number of hoses and hose cabinet for unit pricing. Length of supply line for unit pricing cost of outside connection for fire department connections.

5. Cost of water meters.

6. Water storage tanks where required.

7. Temporary water during construction.

16.2 SPRINKLERS

It is easy to come up with a budget estimate for Sprinkler systems. If the engineering drawing details the system with each head shown, the cost can be estimated by counting the heads and assigning a unit price to each head, say, $125. If no layout is given, most specifications for sprinklers require a sprinkler head every 100 SF, thus taking off the square foot area of the sprinkler area and dividing by 100 will give the number of heads for unit pricing. Extra heads are usually required in stairwells at landings, mechanical rooms, and trash chutes. These can be estimated and the unit price applied.

With a sprinkler system, an alarm system is usually installed. The alarm system is activated by pressure switches installed in the main risers. Depending on the system of alarms, the cost of the pressure switches with their installation is included in the cost of the sprinkler system. The alarm system may be included in the electrical specifications or a separate section with tie-ins to other fire and smoke detection systems.

A checklist of the costs of a sprinkler system is:

1. Count number of heads and assign unit price to each.

2. The branch piping costs on the floors is included in the unit cost of each head. The standpipe system to the floors, if used for both fire hoses and sprinklers, is usually supplied under the plumbing subcontractor's specifications. If a separate sprinkler main is provided, it is provided by the sprinkler subcontractor.

3. Cost for field measurements, sketching and supplying installation, and as-built drawings.

16.3 HEATING, VENTILATION, AND AIR CONDITIONING (HVAC)

Except for construction where individual room air conditioners are used or no heating is required, new construction or major renovations include air conditioning and heating. The general contractor again will rely on a subcontractor to perform this work. In the design of the systems, the architect or engineer tries to budget his designs of the system to be within normal economical range for the building he is designing, for example, $12 to $14 per SF of the building area. The general contractor can thus use the unit price costs developed for the type of building being bid on for his budget estimate.

As stated in Chapter 15, equipment purchased by the general contractor must take into consideration the extra costs of unloading and positioning of equipment required by union contracts.

Since this is a multitrade contract and if the general contractor wishes to act as his own HVAC contractor, here is a list of subcontractors and purchases the general contractor will conceivably be required to assemble to carry out the HVAC section of the contract.

1. Purchase of equipment: boilers, air handlers, air conditioning units, absorption machines, ventilation fans, diffusers, mixing boxes, louvers, prefabricated bases for roof-mounted equipment, tanks, motor controls, motor control centers, many types of pressure gauges, heat sensors, thermostats, radiators, baseboard heating units, individual unit heaters and air conditioning room units, and any special control systems or equipment the architect or engineer may specify such as a computer control system.

2. Subcontractors and services.
   a. Riggers.
   b. Piping or plumbing contractors.
   c. Sheet metal contractor (duct work).
   d. Drafting Service.
   e. HVAC control contractor, vacuum or electrical system.
   f. Electrical control wiring contractor.
   g. A contractor for final balancing of the air conditioning system.
   h. Professional engineer to certify the system.
   i. Air conditioning service contractor.

Since the general contractor's superintendent may not have the background or time to run a HVAC contract, a foreman or an additional superintendent will have to be assigned to carry out the work in the field. Also, an additional person in the office may be required to purchase equipment and to coordinate the subcontractors and shop drawings and follow through on all the paperwork involved in approvals required by the architect for material and equipment. These are additional fixed costs rather than normal overhead for the general contractor. This is why most general contractors rely on the work being performed by a HVAC subcontractor.

16.4 ELECTRICAL

As stated before for the mechanical trades, the general contractor does not attempt to do the electrical work with his own forces. Site work, such as excavation and laying of electrical ducts, foundations for light poles, and precast splice boxes and manholes, may be constructed by the general contractor's own forces. But very few will attempt to do interior electrical work with their own forces.

Estimating the electrical work for budget estimating at the beginning of a bidding period for new construction follows the same procedure as bidding HVAC; a unit price per square foot is assigned to the total square foot area of the building. In general, for renovation work, a fairly close estimate can be achieved by counting the number of light fixtures, receptacles, and switches and assigning them a unit price common to the area.

Electrical drawings are mostly schematic, giving no indication as to where the conduits will be located. Experience is the only way one can estimate with any accuracy the cost of main services and conduit lines to the various areas of the building. This requires considerable guesswork on the part of the estimator. Thus, the general contractor finds it difficult to competitively estimate and do the electrical work with his own forces.

The electrical sections of the specifications cover many services within the building: fire alarm systems, smoke alarms, elevators interacting with the fire alarm system, music and paging systems, HVAC controls, and all pumps, fans, and miscellaneous electrical powered equipment of the plumbing and HVAC contractor. The general contractor may subcontract out to specialty contractors, such as fire alarm systems, instead of one contract for all the electrical work specified under the electrical section of the specifications. Heavy equipment such as emergency electrical generators may be bought directly by the general contractor.

During construction, the electrical contractor is responsible for maintaining temporary light and power. Cost for this can be considerable. Information is available for estimating these costs based on the square foot area of a building and the length of time required to carry out the construction. This is a checklist of electrical costs:

1. Cost of connecting to electrical supply (utility Companies usually bring the service to the inside of the building but panels required for the meter and main disconnect switch are supplied by the electrical contractor).
2. Site work: excavation, manholes, duct conduits, splice boxes, foundations for light poles, and transformers, light poles, and area lighting fixtures.
3. Building service panels.
4. Main service risers.
5. Supply and wiring of all floor and individual area service panels.
6. Supply and installation of all receptacles and switches.
7. Light fixtures, supply and installation.
8. Motor control centers, wiring and supply.
9. Air conditioning control wiring.
10. Wiring of all electrical motor run equipment supplied by other trades.
11. Fire alarm system.
12. Smoke detector system.
13. Paging and music systems.

14. Power to elevator rooms and moving stairs.
15. Temporary light and power during construction.

Cross Reference: Wiley Series of Practical Construction Guides, *Construction Electrical Constructing,* 2nd ed., by J. E. Traister and P. Rosenberg.

A P P E N D I X

Wage Rates (1988–1989) New York City and Vicinity

WAGE DATA: OPERATING ENGINEERS LOCAL #14 NEW YORK CITY

The hourly wage and fringe benefit contributions effective *July 1, 1989* through *June 30, 1990* as follows:

	Wages	+	Fringes	=	Total Package
Wage Rate	$23.88				
Welfare			$ 2.20		
Pension			2.00		
Vacation*			1.05 (taxable)		
Annuity			5.00		
Union Assessment*			0.65 (taxable)		
Training			0.40		
	$23.88	+	$11.30	=	$35.18

*There are two fringe benefit contributions that are *taxable*, namely, Vacation ($1.05) and Union Assessment (65¢). This amount ($1.70) should be added to the hourly rate of $23.88, making the rate for taxable purposes $25.58. After all taxes are deducted, then the amount representing this $1.70 per hour is also deducted from the engineer's net wages because it is included in the stamps you give him with his pay check.

The fringe benefit contributions are to be totally paid in stamps. The Pension and Welfare are paid only for *HOURS WORKED* (does not include premium time). The Vacation, Annuity Union Assessment and Training Program are paid on *HOURS PAID* (includes premium time). When overtime is earned, it necessitates two stamps, one for *ALL HOURS WORKED* ($11.30 per hour) plus one for the premium time earned ($7.10 per hour). For example, if an engineer works 3 hours overtime

345

during a full payroll week, his wage statement should read 38 hours *worked plus* 3 hours premium time. He should receive 41 hours pay in wages, but in stamps he should receive 38 of the $11.30 "S" stamps, plus 3 of the $7.10 "D" stamps).

WAGE DATA: GENERAL BUILDING LABORERS'

The hourly wages and fringe benefits contributions, *effective July 1, 1989 through June 30, 1990* are as follows:

	Wages	+	Fringes	=	Total Package
Wage Rate					
(Includes Dues Check-off)	$18.80				
Welfare Fund			$2.48		
Pension Fund			2.83		
Legal Services Fund					
(Taxable)*			0.15		
Training Program			1.00		
Vacation Fund					
(Taxable)*			1.03		
Annuity Fund			0.30		
Safety Fund			0.05		
	$18.80	+	$7.84	=	$26.64

*Contributions for *Vacation* ($1.03) per hour worked) and *Legal Services Fund* (15¢ per hour worked) *are taxable* and *must be added* to the wage rate of $18.80, bringing the total taxable hourly rate to $19.98, in order to compute and deduct the taxes. After taxes, you are to deduct the *Vacation* and *Legal Services* contributions *and a Dues Check-off of 50¢ per hour worked.*

The Vacation and *Legal Services* Contributions are included when you pay your Fringes of $7.84 to Local 66 Fund Office.

The Dues Check-off of *50¢ per hour worked* is to be paid to the Mason Tenders District Council by *separate check, as you have been doing in the past.*

All Fringes are to be paid on *"hours worked" only* and are not to be paid on "premium hours" earned for overtime work.

WAGE DATA: BRICKLAYERS DISTRICT COUNCIL OF LOCALS 1, 9, 21, 30, 34, 37 AND 41 OF NEW YORK CITY AND LONG ISLAND.

The hourly wage and fringe benefit contributions effective June 1, 1989 through May 31, 1990 are as follows:

	Wages	+	Dues Check-Off and Fringes	=	Total Package
Hourly Wage Rate	24.09				
Dues Check-off*			0.34 (taxable)		
Pension Fund			$3.04		
Insurance and Welfare Fund			2.61		
Annuity Fund† (Journeyman)			3.25		
Industry and Promotion Funds			0.39		
	$24.09	+	$9.63	=	$33.72

*The total contribution for a Dues Check-off (.34¢ per hour × the number of hours worked) is taxable and is to be added to the employee's wages ($24.09 per hour × number of hours worked) in order to compute and deduct the taxes on both. After taxes, the total Dues Check-off is to be deducted from the employee's net wages because it is included when you pay the Dues Check-off and Fringe Benefit contributions at the rate of $9.63 per hour.

†Annuity contributions for Apprentices, Trainees, and Improvers are only $1.70 per hour. Dues Check-off and Fringe Benefit contributions for Apprentices, Trainees, and Improvers therefore are now $8.08 per hour.

BRICKLAYER APPRENTICE WAGE DATA: BRICKLAYERS DISTRICT COUNCIL OF LOCAL 1, 9, 21, 30, 34, 37 AND 41 OF N.Y.C. AND L.I.

Please be advised that effective 6/1/89 thru 5/31/90 the wage rates for Apprentices, Trainees, and Apprentice Improvers (which is a percentage of the journeyman's hourly wage rate based the number of hours each Apprentice, Trainee and Apprentice Improver has worked) are as follows:

Hours of Work	% of Journeymen's Rate	Hourly Rate
1– 750	50%	$12.05
751–1500	60%	14.45
1501–2250	70%	16.86
2251–3000	80%	19.27
3001–3500	90%	21.68
3501–4000	95%	22.89

Fringe Benefit Hourly Contributions effective 6/1/89 thru 5/31/90 for Apprentices, Trainees and Improvers, are as follows:

Dues Check-off*	$0.34 (Taxable)
Pension Fund	3.04

Insurance and Welfare Fund	2.61
Annuity Fund	1.70
Industry and Promotion Funds	0.39
Total Hourly Contribution Rate	$8.08

*The Hourly Wage Rate *plus* the Dues Check-off rate *are taxable* and must be computed in the same manner as the journeymen's wages (see the notice on Bricklayers Wage Data).

WAGE DATA: MASON TENDERS DISTRICT COUNCIL OF NEW YORK CITY

The hourly wages and fringe benefit contributions effective June 1, 1989 through May 31, 1990, are as follows:

	Wages	+	Fringes	=	Total Package
Wage Rate (Includes Dues Check-off)*	$20.67				
Welfare Fund			$2.547		
Pension Fund			1.735		
Annuity Fund			1.25		
Asbestos Training Program			0.05		
Legal Services Fund			0.10		
Industry Fund			0.05		
Political Action Campaign (PAC)*	0.05				
	$20.72		$5.732	=	$26.452

*There are two taxable items included in your Mason Tenders' gross wages for the purpose of computing the withholding taxes. They are the Dues Check-off of 50¢ and the Political Action Campaign contribution of 5¢ per hour worked, making the taxable hourly wage $20.72.

After taxes are deducted from the employee's gross wages, deduct 55¢ for each hour worked (not total hours paid). Pay the Dues Check-off and the PAC by *separate checks, weekly,* to the Mason Tenders District Council.

Classification	Rate per Hour	Supplemental Benefits
Cement and Concrete Workers	$18.65	W.F. $2.2225, P.F. $2.3225, Legal Services Fund $.15, overtime paid at time and a half on weekdays after 7 hour day and Saturdays; double time for Sundays and holidays.
Cement Masons	$15.98 (Plus $1.40 to be allocated between wages and/or supplemental benefits.)	W.F. $2.11, P.F. $2.60, A.F. $4.00, V.F. $2.05, overtime paid at double time after 7 hour day.
Cement Mason Trainee	Wage rate and supplemental benefits not yet determined. Addendum to follow.	
Composition Roofer	$18.68	P.F. $2.05, A.F. $2.55, V.F. $2.40, W.F. $2.00, App.F $.30, Legal Service Fund $.15, overtime paid at time and one half after a 7 hour day.
Roofer Trainee	$11.61	P.F. $1.31, A.F. $1.47, V.F. $1.27, App.F. $.19, W.F. $1.27, Legal Service Fund $.10, overtime paid at double time after 7 hours day.
Derrickman and Riggers (Stone)	$19.36	A.F. $5.25, App.F $.10 P.F. $2.81, V.F. $2.50 W.F. $1.83, overtime paid at double time after 7 hour day.
Diver	$24.52	AJEI $.15, W.F. $2.67, P.F. $2.59, A.F. $1.20, V.F. $1.55, overtime paid at double time in excess of 8 hours per day and all work on Saturdays, Sundays, and Holidays.
Diver Tender	$19.11	

Classification	Rate per Hour	Supplemental Benefits
Dockbuilders/Pile Drivers	$20.66	W.F. $2.67, P.F. $2.59 A.F. $1.20, V.F. $1.55 AJEI $.15, overtime paid at double time after 8 hour day.
Electrician	$24.75	P.F. & W.F. $8\frac{1}{2}$%, D.F. 2% N.E.B.F. 3%, V.&H.F. 11.5%, (effective June 9, 1988 V.&H.F. 10.5%), EF 0.8% A.F.
Electrician	$26.00	$4.00 per day, A.S.B.F. amount equal to employee hourly payroll deduction for FICA (social security); overtime paid at time and one half after 7 hours.
Electrician "M" ("M" rated work shall be defined as jobbing, maintenance, repair work.)	$12.90	Full social security paid by employer, P.F. & W.F. $8\frac{1}{2}$%, EF 0.8%, D.F. 2%, NEBF 3%, AF & ASBF 6.25%, V.F. & H.F. 11.5% (effective June 9, 1988 V&HF 10.5%) paid at time and one half after 8 hour day.
Electrician Trainee 1st Year	$ 7.25	P.F.& W.F. $8\frac{1}{2}$%, D.F. 2% N.E.B.F. 3%, V. & H.F. 11.5% A.S.B.F. $5.00 per
Electrician Trainee 2nd Year	$ 8.45	day (except $2.00 per day for first year
Electrician Trainee 3rd Year	$ 9.45	trainee) overtime paid at time and one
Electrician Trainee 4th Year	$10.45	half after 7 hour day.
Elevator Constructor	$21.97	W.F. $1.945, P.F. $1.55, A.F. $8.00 per day, E.F. $.085, vacation 2 weeks after 1 year.
Elevator Constructor's Helper	$16.48	Overtime paid at the double time rate after 7 hour day.

Engineers (Steel Erection Work)

Maintenance Engineers (Derrick, Travelers, and Climbing Cranes) — $20.34 — AF $4.00, App.F $.25, VF $1.02, WF & PF $2.35, 11 paid holidays, overtime paid after 8 hour day.

Oiler

Truck Crane — $19.07 — "

Crawler Crane — $15.83 — "

Engineers (Building Work)

Maintenance Engineers — $19.88 (Plus 5% increase to be allocated between wages and supplemental benefits.) — "

Installing, repairing, maintaining, dismantling (of all equipment including: Steel Cutting and Bending Machines, Mechanical Heaters, Mine Hoists, Climbing Cranes, Tower Cranes, Linden Peine, Lorain Liebherr, Mannes, or machines of a similar nature, Well Point Systems, Deep Well Pumps, Concrete Mixers with Loading Device, Concrete Plants, Motor Generators when used for temporary power and lights)—driving maintenance trucks and truck-mounted welding machines—all pumps regardless of size and motor power except River Cofferdam Pumps and Well Point Pumps)—when 3 or more motorized concrete buggies (ride type) are utilized on the job site they shall be serviced, maintained and repaired by the

351

Classification	Rate per Hour	Supplemental Benefits
Maintenance Engineer. Maintenance Engineers on Pumps, Generators, Mixers and Heaters	$16.53 (Plus 5% increase to be allocated between wages and supplemental benefits.)	"
Oilers All gasoline, electric, diesel or air operated cranes; Backhoes; Gradalls; Trenching Machines; Concrete Pumps; Gunite Machines; Compressors (3 or more in battery); Overhead Cranes in Power Houses; their duties shall be to assist the Engineer in oiling, greasing and repairing of all machines; Driving Truck Cranes; Driving and Operating Fuel and Grease Trucks.	$18.58 (Plus 5% increase to be allocated between wages and supplemental benefits.)	"
Oilers on Crawler Cranes excluding Crawler Cranes used as Tower Cranes	$15.41 (Plus 5% increase to be allocated between wages and supplemental benefits).	AF $4.00, App.F. $.25, VF $1.02, WF & PF $2.35 11 paid holidays, overtime paid at double time after 8 hour day.
Operator Engineer Trainee	$10.63	"
Operating Engineers Road and Heavy Construction Back Filling Machines, Cranes, Mucking Machines, Dual Drum Pavers	$20.03 (Plus 5% increase to be allocated between wages and supplemental benefits.)	AF $4.50, App.F. $0.30, PF $1.50, VF $1.05, WF $1.55, 11 paid holidays, overtime paid at double time after 8 hour day.
Backhoe, Power Shovel, Hydraulic Clam Shell, Moles and machines of a similar nature.	$21.59 (Plus 5% increase to be allocated between wages and supplemental benefits.)	"

Equipment	Rate	
Boilers (High Pressure), Compressors (Portable—Single or Two in Battery—not over 100' Apart), Pump (River Cofferdom) and Welding Machines (except where Arc is operated by Members of Local 15) Push Button Machines, All Engines, irrespective of Power (Power-Pac) used to drive auxiliary equipment, Air Hydraulic, etc. Compressors (Portable = 3 or more in Battery) Truck Compressor (Engineer Driven), Tugger Machines (Caissons), WellPoint Pumps, Churn Drill.	$13.79 (Plus 5% increase to be allocated between wages and supplemental benefits.)	"
Mixers (Concrete with Loading Attachment), Concrete Pavers Cableways, Land Derricks, Power Houses (low Air Pressure Units)	$14.59 (Plus 5% increase to be allocated between wages and supplemental benefits.)	"
Elevators	$19.46 (Plus 5% increase to be allocated between wages and supplemental benefits.)	"
Gradealls, Keystones, Cranes (with Digging Buckets), Bridge Cranes, Trenching Machines, Vermeer Cutter, and Machines of similar nature	$17.66 (Plus 5% increase to be allocated between wages and supplemental benefits.)	"
Mine Hoist, Cranes, etc., used as Mine Hoist	$20.93 (Plus 5% increase to be allocated between wages and supplemental benefits.)	"
Pile Drivers and Rigs (Employing Dockbuilder Foreman) Derrick Boats,	$21.43 (Plus 5% increase to be allocated between wages and supplemental benefits.)	"
Tunnel Shovels	$20.51 (Plus 5% increase to be allocated between wages and supplemental benefits.)	"

Classification	Rate per Hour	Supplemental Benefits
Concrete Breaking Machines, Hoists (Single Drum), Load Masters, Locomotives and Dinkies over 10 Tons, Hydraulic Crane-Second Engineer.	$18.06 (Plus 5% increase to be allocated between wages and supplemental benefits.)	"
Concrete Pumps, Concrete Plant, Well Drilling Machines, Stone Crushers, Double Drum Hoist, Power Houses (other than above)	$19.01 (Plus 5% increase to be allocated between wages and supplemental benefits.)	"
Concrete Mixer	$18.95 (Plus 5% increase to be allocated between wages and supplemental benefits.)	"
Raise Bore Drill and machines of a similar nature	$20.04 (Plus 5% increase to be allocated between wages and supplemental benefits.)	"
Tower Crane	$26.16 (Plus 5% increase to be allocated between wages and supplemental benefits.)	"
Utility Compressors	$13.70 (Plus 5% increase to be allocated between wages and supplemental benefits.)	"
Off Shift Compressors	$17.19 (Plus 5% increase to be allocated between wages and supplemental benefits.)	"
On Sight Concrete Plant Engineer and On Sight Asphalt Plant Engineer and Vibratory Console	$17.84 (Plus 5% increase to be allocated between wages and supplemental benefits.)	"

Classification	Rate
Operating Engineers/Paving Asphalt Spreaders, Autogrades (CMI), Roto/Mil	$19.46 (Plus 5% increase to be allocated between wages and supplemental benefits.)
Asphalt Roller	$18.95 (Plus 5% increase to be allocated between wages and supplemental benefits.)
Asphalt Plants	$16.79 (Plus 5% increase to be allocated between wages and supplemental benefits.)
Offsight Plant Engineer	$16.66 (Plus 5% increase to be allocated between wages and supplemental benefits.)
Operating Engineers/Concrete Cranes	$21.37 (Plus 5% increase to be allocated between wages and supplemental benefits.)
Operating Engineers/Steel Erection Three Drum Derricks	$23.21 (Plus 5% increase to be allocated between wages and supplemental benefits.)
Cranes, 2 Drum Derricks, Hydraulic Cranes and Fork Lifts	$22.18 (Plus 5% increase to be allocated between wages and supplemental benefits.)
Compressors, Welding Machines	$16.19 (Plus 5% increase to be allocated between wages and supplemental benefits.)

Classification	Rate per Hour	Supplemental Benefits
Operating Engineer/Building Work Hoists, Forklifts, House Cars Plaster (platform machiner), Plaster Bucket, Concrete Pump and all other equipment used for hoisting material.	$20.36 (Plus 5% increase to be allocated between wages and supplemental benefits.)	"
Compressors, Welding Machines (Cutting Concrete-Tank Work), Paint Spraying, Sandblasting, Pumps (with the exclusion of Concrete Pumps), House Car (settlement basis only) All Engines irrespective of Power (Power-Pac) used to drive Auxiliary Equipment, Air, Hydraulic, etc. Boilers.	$16.77 (Plus 5% increase to be allocated between wages and supplemental benefits.)	"
Double Drum	$21.57 (Plus 5% increase to be allocated between wages and supplemental benefits.)	"
Stone Derrick, Cranes, Hydraulic Cranes	$21.16 (Plus 5% increase to be allocated between wages and supplemental benefits.)	"
Glaziers	$19.80	AF $4.75, PF $1.26, Vacation & Holiday Fund $1.57, WF $1.26, overtime paid at double time after 7 hour day.
Glazier Trainee	Wage rate and supplemental benefits not yet determined. Addendum to follow.	
Heat and Frost Insulator	$20.17	WF $2.44, PF $1.56, VF $2.50, AF $5.00, overtime paid at double time after 7 hour day.

Classification	Rate	Benefits
Highway and Paving workers		WF $2.15, PF $2.25, AF $1.50, 7 paid holidays, overtime paid at time and one half after 8 hour day.
Asphalt Raker	$18.19	
Tamper	$17.23	
Foremen & Form Setter	$19.08	
Laborer (Highway and Paving)	$16.85	
All Other Persons	$17.18	
Screedman	$18.47	
Iron Workers (Ornamental)	$17.40	VF $2.50, PF $2.64, WF $2.39, AF $4.85, App.F. $0.25, overtime paid at double time after 7 hour day.
Ornamental Iron Worker Trainee		Wage rate and supplemental benefits not yet determined. Addendum to follow.
Iron Worker (Structural)	$18.00	Topping Out Fund $2.00, PF $3.30, WF $2.00, VF $4.20, AF $5.50, App.F. $0.28, overtime paid at time and one half after 8 hour day.
Structural Iron Worker Trainee		Wage rate and supplemental benefits not yet determined. Addendum to follow.
Laborers (Building, Concrete, Excavating, and Common)	$15.75	AF $2.05, PF $1.35, WF $3.38, overtime paid at time and a half after 8 hour day, 8 paid holidays.
Linoleum Layer (Asphalt Tile Carpet)	$20.66	WF $2.67, PF $2.59, AF $1.20, AJEI $0.15, VF $1.55 overtime paid at time and one half after 7 hour day. Collective Bargaining Agreement expired June 30, 1987. Addendum to follow

Classification	Rate per Hour	Supplemental Benefits
Marble Carvers	$17.75	WF $.96, PF $1.18, VF $3.50, AF $4.30, Sup.F. $0.40, tool allowance $0.50 per hour travel expense $10.00 per day, overtime paid at double time after 7 hour day.
Marble Helper	$15.27	WF $.96, PF $1.18, VF $3.50, AF $4.30, Sup.F $0.40, tool allowance $0.50 per hour, travel expense $10.00 per day, overtime paid at double time after a 7 hour day.
Metallic Lathers	$20.61	WF $1.745, PF $.985, App.F. $0.80, VF $2.00, AF $3.50, overtime paid at double time after 7 hour day.
Metallic Lather Trainee		
1st year	$8.08	"
2nd year	75% of metallic lathers' wage	"
3rd year	85% of metallic lathers' wage	"
Millwright	$19.29	AJEI $.15, WF $2.67, PF $2.59, AF $2.94, VF $2.40, overtime paid at double time after 7 hour day.
Mosaic & Terrazzo Worker	$18.10 (Effective 7/1/86 $0.70 to be allocated between wages and supplemental benefits and effective 1/1/87 $0.71 to be allocated between wages and supplemental benefits.)	WF $1.48, PF $2.90, VF $1.50, and overtime paid at time and one half after 7 hour day; double time on Saturday.

Mosaic & Terrazzo Helper — $17.61 (Effective 7/1/86 $0.79 to be allocated between wages and supplemental benefits and effective 1/1/87 $0.80 to be allocated between wages and supplemental benefits.) WF $1.48, PF $2.90, VF $0.50, and overtime paid at time and a one half after 7 hour day; double time on Saturday, Sunday, and holidays.

Machine Operator Grinder — $17.61 (Effective 7/1/86 $0.79 to be allocated between wages and supplemental benefits and effective 1/1/87 $0.80 to be allocated between wages and supplemental benefits.)

Painters
- Brush & Roller — $18.00
- Spray & Scaffold — $21.86
- Fire Escape — $20.57
- Decorative — $20.57
- Painter Trainee — Wage rate and supplemental benefits not yet determined. Addendum to follow.

WF 9.5%, PF 9.5%, AF 9%, VF 6%, overtime paid at time and one half after 7 hour day.

Paperhanger — $23.75

Painters on Structural Steel — $22.00

WF 9.5%, PF 14%, AF 10%, VF 6%

Plumbers (Manhattan and Bronx) — $23.16 (Plus $1.25 to be allocated between wages and/or supplemental benefits. Effective 1-6-88 Plus $1.25 to be allocated between wages and/or supplemental benefits.)

WF 17%, VF 6%, PF $1.10, AF 11%, App.F. $\frac{1}{2}$ of 1% Travel Expense $.050, overtime paid at double time after a 7 hour day except time and one half on Saturday. Holiday Fund $1.505, VF $1.605, PF $3.15, WF $2.67, EF $0.01, National Pension Fund $0.80, ASBF $6.00 per day, Additional ASBF $2.00, Safety Expense $0.28, overtime paid at double time after 7 hour day.

Classification	Rate per Hour	Supplemental Benefits
Plumber Trainee		ASBF $3.00 per day, VF $0.65, Holiday Fund $0.428, WF $2.392, EF $0.01, National Pension Fund $0.25, Safety Expenses $0.14, overtime paid at double time after 7 hour day.
1st year 1st Grade	$ 6.05	
2nd Grade	$ 6.55	
2nd year 3rd Grade	$ 7.30	
4th Grade	$ 8.05	
3rd year 5th Grade	$ 8.80	
6th Grade	$ 9.55	
4th year 7th Grade	$10.30	
8th Grade	$11.05	
5th year 9th Grade	$11.80	
10th Grade	$12.55	
Plumbers Repair and Maintenance Jobbing Work (Manhattan and Bronx) (Work defined as jobbing shall be any repair and/or replacement of the present plumbing system that does not change existing roughing.)	$15.50 (Plus $.67 to be allocated between wages and/or supplemental benefits. Effective 1-6-88 $.66 to be allocated between wages and/or supplemental benefits.)	National Pension Fund $0.25, EF $0.01, ASBF $2.00 per day, Safety Expense $0.14, Holiday Fund $0.70, VF $1.02, WF $2.96, overtime paid at time and a half after 8 hour day.
Plumbers (Brooklyn and Queens)	$21.86	WF $1.79, PF $2.14, VF $1.25, HF $1.50, Trade Education Fund $0.22, Additional ASB $1.25, ASB (Daily) $6.00, National Pension Fund $1.10, Safety Expense $0.28, overtime paid at time and one half after 7 hour day.
Plumber Trainee		Trade Education Fund $0.42, Safety Expense $0.28, overtime paid at time and one half after 7 hour day.
1st year 1st Grade	$ 5.82	

	2nd Grade	$ 5.82	WF $0.55, Safety Expenses $0.28, overtime paid at time and one half after 7 hour day.
2nd year	3rd Grade	$ 6.56	Additional ASB $0.63, PF $1.09, ASB (Daily) $3.00, WF $0.92, National Pension Fund $0.54, Safety Expense $0.28, VF $0.63, HF $0.75, Trade Education $0.12, overtime paid at time and one after 7 hour day.
	4th Grade	$ 7.65	
3rd year	5th Grade	$ 8.74	
	6th Grade	$11.20	
4th year	7th Grade	$11.40	
	8th Grade	$15.30	
Plumbers (Staten Island)		$26.12 (Plus $1.17 to be allocated between wages and/or supplemental benefits.)	PF $3.08, WF $2.35, App.F. $0.35, ASBF $7.20 per day, overtime paid at double time for work after a 7 hour day.
Plumber Jobbing & Alteration Mechanic (Staten Island)		$16.23	PF $2.00, WF $2.35, App.F. $0.13, overtime paid at time and one half after a 8 hour day.
Teamsters			
Automobile Chauffeur		$15.825	WF $3.00, PF $2.8525, AF $3.3125, ($3.00 for redimix drivers). Vacation 1 day for every 15 days worked. Sick Days 1 day for every 30 days worked (for redi-mix drivers only). 11 PH. Overtime paid at time and one half after 8 hour day.
Heavy Equipment Trailer Driver		$16.575	
Euclid & Turnapull Operator		$16.225	
Six Wheeler (3 Axle) Tractors & Trailers		$16.325	
Boom Truck Driver		$16.325	
Redi-Mix Driver (Sand & Gravel)		$16.00	
Tile Layer		$20.66	WF $2.67, PF $2.59, VF $1.55, AJEI $0.15, AF $1.20, overtime paid at time and one half after 7 hour day.

Classification	Rate per Hour	Supplemental Benefits
Plumbers Laborer/Pipe Layer (Cast Iron, Steel Sewer)	$15.75	AF $2.05, PF $1.35, WF $3.38, 8 PH, overtime paid at time and one half after eight hour day.
Pointers (Waterproofers) and Caulker Mechanics (Building Renovation Exterior)	$21.34	WF $1.75, PF $3.00, AF $2.25, overtime paid at double time after a 7 hour day.
Sandblaster	$21.59	"
Steamblaster (et al.)	$21.94	"
Sheet Metal Workers	$23.35	VF $1.501, WF $2.152, PF $2.08, AF $2.65, Supp. Employment F. $0.11, Educational Trust F. $0.12, overtime paid at double time for work after a 7 hour day.
Steamfitter	$23.00	VF $2.50, Sec. Ben. Fund $2.50, PF $1.60, WF $2.05, EF $0.17, overtime paid at double time after a 7 hour day.
Steamfitter Trainee	$17.25	VF $1.88, Sec. Ben. Fund $2.50, PF $1.20, WF $1.54, EF $0.13, overtime paid at after a 7 hour day.
Stone Setter	$20.86	WF $1.75, PF $2.00, AF $4.50, VF $2.00, ASBF $0.60, App.F. $0.10, tool allowance $3.00 per day, overtime paid at double time after 7 hour day.

Glossary

Bank yard. A cubic yard of earth measured in the cut. A cut yard.

Collection hopper. A small hopper used to feed concrete into a wall form or into an elephant trunk.

Cut. The area where excavation is being made, or the volume of excavation, measured in bank or cut yards. The depth of excavation. A decrease in a cost or in a price.

Deadman. A ground anchorage, usually used in connection with a wire rope bracing system.

Elephant trunk. A tube 8 or 10 inches in diameter used to place concrete when the height of fall is greater than about 6 feet. The tube may be made of canvas, rubber, or metal. Elephant trunks are frequently but incorrectly called tremie tubes, or tremies.

Fill. The area where embankment is being made, or the volume of embankment measured in compacted cubic yards. The height of embankment.

Furring. A system of light wood framing used to enclose ductwork, to lower a portion of a ceiling, or to increase the thickness of a wall for architectural reasons.

Grizzly. A heavy-duty grid made of steel shapes or plates used to screen out oversize material in earthwork or quarrying operations.

Loose yard. A cubic yard of earth measured after the material has been excavated. Freshly stockpiled earth and all stockpiled aggregates are normally considered to be in the loose state.

Neat excavation. Excavation made to the neat concrete lines, which are vertical planes at the outermost edges of the concrete.

Sacked finish. A surface finish used to improve the appearance of concrete. The surface is wet thoroughly, and the free moisture on the surface is allowed to dry. Mortar having the consistency of thick cream is applied to the surface with a rubber float. The mortar consist of about two-thirds part portland cement, one-third part white cement, and two parts of well-graded minus 30 mesh sand. The mortar is rubbed over the surface with burlap pads until all air and water pockets are filled. The surface is then dusted with the dry mortar mix and rubbed again with burlap to achieve a uniform color. On completion, no mortar will appear on the surface, but all voids will be filled. The completed surface is moisture cured for about 3 days.

Scaffold jacks. Prefabricated metal kneebraces used in conection with temporary staging on wall forms.

Soffit. The undersurface of a slab, a beam, or an arch.

Spandrel beam. A beam, at the edge of a concrete slab, that extends both above and below the slab.

Spandrel well. The building wall that extends from the sill of an upper window to the head of the window immediately below.

Starter wall. A portion of concrete wall, usually $1\frac{1}{2}$ inches in height, placed as a part of the wall foundation and subsequently used to spread and align the wall forms.

Tremie tube. A steel pipe 10 or 12 inches in diameter with a collection hopper at the upper end used to place concrete under water.

Wales. Wooden beams normally used in pairs to support the wall form studs. Wales are normally placed at right angles to the studs and are usually made of material the same size as the studs. The wales on either side of a formed wall are tied together, through the wall, with form ties.

Cross Reference: For more complete information, refer to *Construction Glossary*, by J. S. Stein, a volume in this series.

About the Author

Leo Diamant received his B.C.E. from Cornell University in 1947. He then spent ten years as a structural engineer designing facilities for power plants, underground pumping stations and a wide variety of industrial buildings. He became the chief strctural designer for a company whose principal business was the manufacture of rolling mills, forging presses, and extrusion presses. The company specialized in turnkey jobs, thus requiring the buildings, special equipment foundations, and all support facilities required by the client for a complete operating manufacturing facility.

He was involved in the initial stages of missile development at Cape Canaveral, Florida. He designed and also supervised the construction of launching platforms at the Cape. He received special commendations for his work from the Navy and missile contractors for meeting impossible schedules.

Always wanting to be involved directly in construction, he left the engineering office and joined a general contracting firm in New York, as a construction superintendent. He supervised the construction of various hospitals, apartment buildings, and the renovations of a number of industrial properties. After spending ten years as a superintendent, he came back to the office as an estimator and project manager for general contractors.

In July 1987, he retired from his position as chief estimator and project manager from Bristol Construction Company, a general contractor involved in building construction for both public works and private developers.

He continues to be active in retirement, teaching estimating at New York University School of Continuing Education, doing some private estimating for clients, and serving as an arbitrator for the construction industry with the American Arbitration Association.

INDEX

Estimating process, 3
Estimator, job superintendent and, 4, 5
Excavation:
earth, 59, 60
hand, 80, 81
roadway, 90, 91
structure, costs of, 80
Expansion joint armor, precast concrete girder
bridge decks, cost of, 259
Expansion joint material, cost of, 181
Exterior walls, masonry, 308
architectural designed block, 308
block backup, 308
brick veneer, 308
cavity, 308
metal stud backup, 308
solid brick, 308
wood stud backup, 308

Falsework design, cost estimate of, 173–175
Falsework footings, poured-in-place, unit labor
cost, 270
Falsework lumber, costs of, 173
Federal Social Security taxes, 41
Federal Unemployment tax, 41
Fences, cost of, 37, 162
Field experience, 1
Finish carpentry, cost estimate, 319, 320
Finishing, additional, 179
Finishing machines, rental charges for, 181
Finishing roadway, 142, 143
Finishing trades, 332–337
ceilings, 334–336
drywall, 332–334
flooring, 336–337
painting, 337
Fire insurance, 36
Fixed price contracts, 16
Flat slab bridges:
completion, total time required for, 203, 204
concrete take-off, 195, 197
cost estimate of, 195
metal beam bridge rail, estimated cost of, 199
structure concrete estimate, 197, 199
Flooring, 336–337
additional costs, 336–337
carpeting, 336
checklist for estimating, 337
flash patching, 337
preparation, 336, 337
resilient, 336
take off, 336
underlayment, 337
waste, 336

wood, 337
parquet, 337
strip, 337
Floors, beam and slab, 286–289
concrete joist, 390–394
Foreign currencies, converting unit costs to, 57, 58
Form lumber, costs of, 173, 285, 286, 290, 294, 301
Form making strips, cost of, 178, 179
Form oil, cost of, 178
Form ties, cost of, 177
Formwork, cost of, 172, 178, 179
Foundations, concrete building, cost of, 277
concrete take-off, 277–280
unit costs, 280–281
Fringe benefits, 40

General costs, concrete building work, 272–277
General job costs, 33–47
General overhead, cost estimate of, 45–47
Girder bridge decks, precast concrete, unit costs, 252, 259–264
Girder bulkheads, prestressed box girder
bridges, cost of, 248
Girder erection, cost of, 252–253
Girders, cleaning, cost of, 182
precast concrete, cost of, 249–251
Grade beams, 280–281
Grade slabs, unit costs, 303–306
Grouting, broken concrete riprap, cost of, 164
Grubbing, costs of, 36, 142, 143, 193
Guard rail, metal beam, cost of, 162

Hand excavation, costs of, 80, 81
Hand tools, cost of, 178
electric, rental charge, 178
Hardware, 329, 330
designation on plan, 329
installation costs, 329, 330
manufacturers' catalog, 330
purchase, 329
schedule, 330
sets, 329
specifications, 329
suppliers, 329
take off, 329
Haul distances, imported borrow, cost of, 125
roadway excavation, cost of, 95–104
Haul road, roadway excavation, maintaining, 120–123
Headwalls, pipe, unit costs, 269
Heating, cost of, 37